ELECTRICITY
with bulbs, batteries, foil, clothespins and simple things

SCIENCE WITH SIMPLE THINGS SERIES

Conceived and written by
RON MARSON

Illustrated by
PEG MARSON

342 S Plumas Street
Willows, CA 95988

www.topscience.org

TOPS LEARNING SYSTEMS

WHAT CAN YOU COPY?

Dear Educator,

　　Please honor our copyright restrictions. We offer liberal options and guidelines below with the intention of balancing your needs with ours. When you buy these labs and use them for your own teaching, you sustain our work. If you "loan" or circulate copies to others without compensating TOPS, you squeeze us financially, and make it harder for our small non-profit to survive. Our well-being rests in your hands. Please help us keep our low-cost, creative lessons available to students everywhere. Thank you!

PURCHASE, ROYALTY and LICENSE OPTIONS

TEACHERS, HOMESCHOOLERS, LIBRARIES:

　　We do all we can to keep our prices low. Like any business, we have ongoing expenses to meet. We trust our users to observe the terms of our copyright restrictions. While we prefer that all users purchase their own TOPS labs, we accept that real-life situations sometimes call for flexibility.

　　Reselling, trading, or loaning our materials is prohibited unless one or both parties contribute an Honor System Royalty as fair compensation for value received. We suggest the following amounts – let your conscience be your guide.

　　HONOR SYSTEM ROYALTIES: If making copies from a library, or sharing copies with colleagues, please calculate their value at 50 cents per lesson, or 25 cents for homeschoolers. This contribution may be made at our website or by mail (addresses at the bottom of this page). Any additional tax-deductible contributions to make our ongoing work possible will be accepted gratefully and used well.

　　Please follow through promptly on your good intentions. Stay legal, and do the right thing.

SCHOOLS, DISTRICTS, and HOMESCHOOL CO-OPS:

PURCHASE Option: Order a book in quantities equal to the number of target classrooms or homes, and receive quantity discounts. If you order 5 books or downloads, for example, then you have unrestricted use of this curriculum for any 5 classrooms or families per year for the life of your institution or co-op.

　　2-9 copies of any title: 90% of current catalog price + shipping.

　　10+ copies of any title: 80% of current catalog price + shipping.

ROYALTY/LICENSE Option: Purchase just one book or download *plus* photocopy or printing rights for a designated number of classrooms or families. If you pay for 5 additional Licenses, for example, then you have purchased reproduction rights for an entire book or download edition for any **6** classrooms or families per year for the life of your institution or co-op.

　　1-9 Licenses: 70% of current catalog price per designated classroom or home.

　　10+ Licenses: 60% of current catalog price per designated classroom or home.

WORKSHOPS and TEACHER TRAINING PROGRAMS:

　　We are grateful to all of you who spread the word about TOPS. Please limit copies to only those lessons you will be using, and collect all copyrighted materials afterward. No take-home copies, please. Copies of copies are strictly prohibited.

　　For licensing, honor system royalty payments, contact: **www.TOPScience.org**; or **TOPS Learning Systems 342 S Plumas St, Willows CA 95988**; or inquire at **customerservice@topscience.org**

ISBN 978 - 0 - 941008 - 53 - 2

CONTENTS

PREPARATION AND SUPPORT

ACTIVITIES AND LESSON NOTES

SUPPLEMENTARY PAGES

A TOPS Teaching Model

If science were only a set of explanations and a collection of facts, you could teach it with blackboard and chalk. You could require students to read chapters in a textbook, assign questions at the end of each chapter, and set periodic written exams to determine what they remember. Science is traditionally taught in this manner. Everybody studies the same information at the same time. Class togetherness is preserved.

But science is more than this. It is also process — a dynamic interaction of rational inquiry and creative play. Scientists probe, poke, handle, observe, question, think up theories, test ideas, jump to conclusions, make mistakes, revise, synthesize, communicate, disagree and discover. Students can understand science as process only if they are free to think and act like scientists, in a classroom that recognizes and honors individual differences.

Science is both a traditional body of knowledge and an individualized process of creative inquiry. Science as process cannot ignore tradition. We stand on the shoulders of those who have gone before. If each generation reinvents the wheel, there is no time to discover the stars. Nor can traditional science continue to evolve and redefine itself without process. Science without this cutting edge of discovery is a static, dead thing.

Here is a teaching model that combines both the content and process of science into an integrated whole. This model, like any scientific theory, must give way over time to new and better ideas. We challenge you to incorporate this TOPS model into your own teaching practice. Change it and make it better so it works for you.

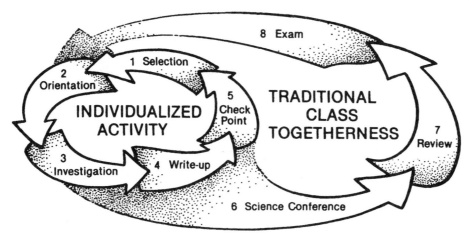

1. SELECTION

Students generally select activity pages in sequence, because new concepts build on old ones in a specific order. There are, however, exceptions to this pattern: students might skip a lesson that is not challenging; repeat an activity with doubtful results; add an experiment to answer their own "what-would-happen-if?" questions.

Working at their own pace, students fall into a natural routine that creates stability and order. They still have questions and problems, to be sure, but remain purposefully engaged with a definite sense of direction.

2. ORIENTATION

Any student with basic reading skills can successfully interpret our carefully designed activity page directions. If your class is new to TOPS, it may take a while for your students to get used to following directions by themselves, and to trust in their own problem-solving ability.

When students ask you for help, first ask them to read what they don't understand. If they didn't read the instruction in the first place, this should clear things up. Identify poor readers in your class. Whey they ask, "What does this mean?" they may be asking in reality, "Will you please read these directions aloud?"

Beyond reading comprehension, certain basic concepts and skills may also be necessary to complete some activity sheets. You can't, for example, expect students to measure the length of something unless they know how to use a ruler as well. Anticipate and teach prerequisite concepts and skills (if any) at the beginning of each class period, before students begin their daily individualized work. Different age groups will require different levels of assistance: primary students will need more introductory support than middle school students; secondary students may require none at all.

3. INVESTIGATION

Students work through the activity pages independently and cooperatively, They follow their own experimental strategies and help each other. Encourage this behavior by helping students only after they have tried to help themselves. As a resource teacher, you work to stay out of the center of attention, responding to student questions rather than posing teacher questions.

Some students will progress more rapidly than others. To finish as a cohesive group, announce well in advance when individualized study will end. Expect to generate a frenzy of activity as students rush to meet your deadline. While slower students finish those core activities you specify, challenge your more advanced students with Extension activities, or to design original experiments.

4. WRITE-UP

Activity pages ask students to explain the how and why of things. Answers may be brief and to the point, with the exception of those that require creative writing. Students may accelerate their pace by completing these reports out of class.

Students may work alone, or in cooperative lab groups. But each one should prepare an original write-up, and bring it to you for approval. Avoid an avalanche of write-ups near the end of the unit by enforcing this simple rule: each write-up must be approved before starting the next activity.

5. CHECK POINT

Student and teacher together evaluate each write-up on a pass/no-pass basis. Thus no time is wasted haggling over grades. If the student has made reasonable effort consistent with individual ability, check off the completed activity on a progress chart. Students keep these in notebooks or assignment folders kept on file in class.

Because the student is present when you evaluate, feedback is immediate and effective. A few moments of your personal attention is surely more effective than tedious margin notes that students may not heed or understand. Remember, you don't have to point out every error. Zero in on particular weaknesses. If reasonable effort is not evident, direct students to make specific improvements and return for a final check.

A responsible lab assistant can double the amount of individual attention each student receives. If he or she is mature and respected by your students, have the assistant check even-numbered write-ups, while you check the odd ones. This will balance the work load and assure equal treatment.

6. SCIENCE CONFERENCE

Individualized study has ended. This is a time for students to come together, to discuss experimental results, to debate and draw conclusions. Slower students learn about the enrichment activities of faster classmates. Those who did original investigations or made unusual discoveries share this information with their peers, just like scientists at a real conference.

This conference is an opportunity to expand ideas, explore relevancy and integrate subject areas. Consider bringing in films, newspaper articles and community speakers. It's a meaningful time to investigate the technological and social implications of the topic you are studying. Make it an event to remember.

7. REVIEW

Does your school have an adopted science textbook? Do parts of your science syllabus still need to be covered? Now is the time to integrate traditional science resources into your overall program. Your students already share a common background of hands-on lab work. With this base of experience, they can now read the text with greater understanding, think and problem-solve more successfully, communicate more effectively.

You might spend just a day here, or an entire week. Finish with a review of major concepts in preparation for the final exam. Our review/test questions provide an excellent resource for discussion and study.

8. EXAM

Use any combination of our review/test questions, plus questions of your own, to determine how well students have mastered the concepts they've been learning.

Now that your class has completed a major TOPS learning cycle, it's time to start fresh with a brand new topic. Those who messed up and got behind don't need to stay there. Everyone begins the new topic on an equal footing. This frequent change of pace encourages your students to work hard, to enjoy what they learn, and thereby grow in scientific literacy.

B

Getting Ready

Here is a checklist of things to think about and preparations to make before beginning your first lesson on ELECTRICITY.

✔ Review the scope and sequence.

Take just a few minutes, right now, to page through all 20 lessons. Pause to read each *Objective* (top left column of the Teaching Notes) and scan each lesson.

✔ Set aside appropriate class time.

Allow an average of perhaps 1 class period per lesson (more for younger students), plus time at the end of this module for discussion, review and testing. If you teach science every day, this module will likely engage your class for about 4 weeks. If your schedule doesn't allow this much science, consult the logic tree on page E to see which activities you can safely omit without breaking conceptual links between lessons.

✔ Number your activity sheet masters.

The small number printed in the top right corner of each activity page shows its position within the series. If this ordering fits your schedule, copy each number into the blank parentheses next to it. Use pencil, as you may decide to revise, rearrange, add or omit lessons the next time you teach this topic. Insert your own better ideas wherever they fit best, and renumber the sequence. This allows your curriculum to adapt and grow as you do.

✔ Photocopy sets of student activity sheets.

Supply 1 per student, plus supplementary pages as required. Store these in manila folders for convenient access by your students. Please honor our copyright notice at the front of this book. We allow you, the purchaser, to photocopy all permissible materials, as long as you limit the distribution of copies you make to the students you personally teach. Encourage other teachers who want to use this module to purchase their own books. This supports TOPS financially, enabling us to continue publishing new titles for you.

✔ Collect needed materials.

See page D, opposite, for details.

✔ Organize a way to track assignments.

Keep student work on file in class. If you lack a file cabinet, a box with a brick will serve. File folders or notebooks both make suitable assignment organizers. Students will feel a sense of accomplishment as they see their folders grow heavy, or their notebooks fill, with completed assignments. Since all papers stay together, reference and review are facilitated.

Ask students to number a sheet of paper from 1 to 20 and tape it inside the front cover of their folders or notebooks. Track individual progress by initialing lesson numbers as daily assignments pass your check point.

✔ Review safety procedures.

In our litigation-conscious society, we find that publishers are often more committed to protecting themselves from liability suits than protecting students from physical hazards. Lab instructions are too often filled with spurious advisories, cautions and warnings that desensitize students to safety in general. If we cry "Wolf!" too often, real warnings of present danger may go unheeded.

At TOPS we endeavor to use good sense in deciding what students already know (don't stab yourself in the eye) and what they should be told (don't look directly at the sun.) Pointed scissors, pins and such are certainly dangerous in the hands of unsupervised children. Nor can this curriculum anticipate irresponsible behavior or negligence. As the teacher, it is ultimately your responsibility to see that common-sense safety rules are followed; it is your students' responsibility to respect and protect themselves and each other.

✔ Communicate your grading expectations.

Whatever your grading philosophy, your students need to understand how they will be assessed. Here is a scheme that counts individual effort, attitude and overall achievement. We think these three components deserve equal weight:

• Pace (effort): Tally the number of check points and extra credit experiments you have initialed for each student. Low ability students should be able to keep pace with gifted students, since write-ups are evaluated relative to individual performance standards on a pass/no-pass basis. Students with absences, or those who tend to work slowly, might assign themselves more homework out of class.

• Participation (attitude): This is a subjective grade, assigned to measure personal initiative and responsibility. Active participators who work to capacity receive high marks. Inactive onlookers who waste time in class and copy the results of others receive low marks.

• Exam (achievement): Activities point toward generalizations that provide a basis for hypothesizing and predicting. The Review/Test questions beginning on page G will help you assess whether students understand relevant theory and can apply it in a predictive way.

Gathering Materials

Listed below is everything you'll need to teach this module. Buy what you don't already have from your local supermarket, drugstore or hardware store. Ask students to bring recycled materials from home.

Keep this classification key in mind as you review what's needed.

general on-the-shelf materials: Normal type suggests that these materials are used often. Keep these basics on shelves or in drawers that are readily accessible to your students. The next TOPS module you teach will likely utilize many of these same materials.	**special in-a-box materials:** *Italic type suggests that these materials are unusual. Keep these specialty items in a separate box. After you finish teaching this module, label the box for storage and put it away, ready to use again.*
(substituted materials): Parentheses enclosing any item suggests a ready substitute. These alternatives may work just as well as the original. Don't be afraid to improvise, to make do with what you have.	***optional materials:** An asterisk sets these items apart. They are nice to have, but you can easily live without them. They are probably not worth an extra trip to the store, unless you are gathering other materials as well.

Everything is listed in order of first use. Start gathering at the top of this list and work down. Ask students to bring recycled items from home. The Teaching Notes may occasionally suggest additional *Extensions*. Materials for these optional experiments are listed neither here nor under *Materials*. Read the extension itself to determine what new items, if any, are required.

Quantities depend on how many students you have, how you organize them into activity groups, and how you teach. Decide which of these 3 estimates best applies to you, then adjust quantities up or down as necessary:

$Q_1/Q_2/Q_3$

- **Single Student:** Enough for 1 student to do all the experiments.
- **Individualized Approach:** Enough for 30 students informally working in pairs, all self-paced.
- **Traditional Approach:** Enough for 30 students, organized into pairs, all doing the same lesson.

KEY:	*special in-a-box materials* *(substituted materials)*	general on-the-shelf materials *optional materials

$Q_1/Q_2/Q_3$

1 roll	masking tape, 3/4 inch wide – see teaching notes 1 for possible substitutions
1 box	aluminum foil, light duty OK
1 / 15 / 15	pairs of scissors
3 / 30 / 30	dry cells, size D, 1.5 volt
2 / 30 / 30	flashlight bulbs, 4.5 volt, designed for 3 size D dry cells, often marked PR 3. Bulbs with a "collar" (a protruding rim of metal below the bulb) will work better than screw-in kinds. It is important to use bulbs from the same manufacturer with the same resistance rating so they will shine equally bright See Teaching Notes 3 for an economical alternative.
1 box	paper clips
15 / 150 / 150	rubber bands: narrow, medium size are easiest to use
10 / 100 / 150	pennies
5 / 150 / 150	index cards, 3 x 5 inch preferable
1 / 2 / 3	paper punch tools
6 / 110 / 130	wooden spring-action clothespins
1 box	steel wool, unsoaped, fine grade, about as thick as human hair
1 / 10 / 30	straight pins: 1-inch steel pins
3 / 30 / 30	balloons, any size

Sequencing Activities

This logic tree shows how all the activities in this book tie together. In general, students begin at the trunk of the tree and work up through the related branches. Lower level activities support the ones above.

You may, at your discretion, omit certain activities or change their sequence to meet specific class needs. However, when leaves open vertically into each other, those below logically precede those above, and should not be omitted.

When possible, students should complete the activities in the same sequence as numbered. If time is short, however, or certain students need to catch up, you can use this logic tree to identify concept-related horizontal activities. Some of these might be omitted, since they serve to reinforce learned concepts rather than introduce new ones.

For whatever reason, when you wish to make sequence changes, you'll find this logic tree a valuable reference. Parentheses in the upper right corner of each activity page allow you total flexibility. They are blank so you can pencil in sequence numbers of your own choosing.

ELECTRICITY 32
E

Gaining a Whole Perspective

Science is an interconnected fabric of ideas woven into broad and harmonious patterns. Use extension ideas in the teaching notes plus the outline presented below to help your students grasp the big ideas — to appreciate the fabric of science as a unified whole.

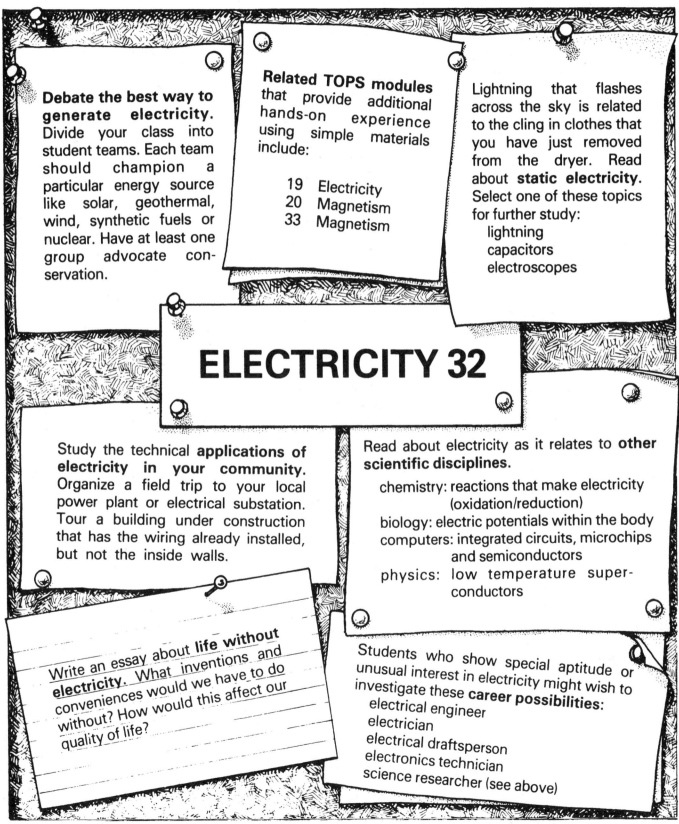

Debate the best way to generate electricity. Divide your class into student teams. Each team should champion a particular energy source like solar, geothermal, wind, synthetic fuels or nuclear. Have at least one group advocate conservation.

Related TOPS modules that provide additional hands-on experience using simple materials include:

19 Electricity
20 Magnetism
33 Magnetism

Lightning that flashes across the sky is related to the cling in clothes that you have just removed from the dryer. Read about **static electricity**. Select one of these topics for further study:
lightning
capacitors
electroscopes

ELECTRICITY 32

Study the technical **applications of electricity in your community.** Organize a field trip to your local power plant or electrical substation. Tour a building under construction that has the wiring already installed, but not the inside walls.

Read about electricity as it relates to **other scientific disciplines.**

chemistry: reactions that make electricity (oxidation/reduction)
biology: electric potentials within the body
computers: integrated circuits, microchips and semiconductors
physics: low temperature super-conductors

Write an essay about **life without electricity.** What inventions and conveniences would we have to do without? How would this affect our quality of life?

Students who show special aptitude or unusual interest in electricity might wish to investigate these **career possibilities:**
electrical engineer
electrician
electrical draftsperson
electronics technician
science researcher (see above)

Review / Test Questions

Photocopy these test questions. Cut out those you wish to use, and tape them onto white paper. Include questions of your own design, as well. Crowd them all onto a single page for students to answer on their own papers, or leave space for student responses after each question, as you wish. Duplicate a class set, and your custom-made test is ready to use. Use leftover questions as a class review in preparation for the final exam.

activity 1
Draw a way to light a bulb with a dry cell. Draw another way that doesn't work.

activity 2
Connect bulbs and dry cells with lines to show how to light each bulb.

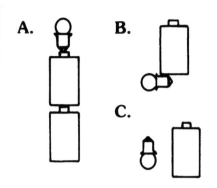

activity 3
Predict if these bulbs will light. Give a reason for each prediction.

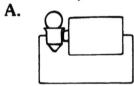

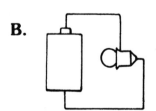

activity 4
Circle one group (A or B) to show which cells make a bulb shine brighter. Do this in all three rows of problems.

Group A or Group B?

activity 5
Number these 5 groups of cells by how bright they make the bulb shine. Write *1* in the blank next to the brightest, *2* for the next brightest, and so on.

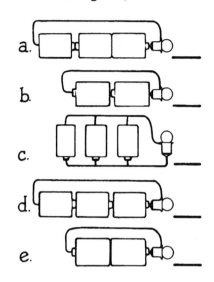

activity 6
You have discovered a UFO (unidentified fallen object) in your back yard. You want to find out if it's a conductor or insulator of electricity. Use words and pictures to tell what you would do.

activity 7

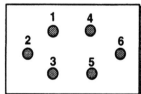

a. A bulb lights with holes 1, 4 and holes 4, 5. Therefore the bulb *must* light with hole(s) _____ as well.

b. Hole 1 lights only with 2. Hole 2 therefore *cannot* light with holes _____.

activity 8a
Which end of the dry cell is the positive pole? When you build a circuit, in what direction do the electrons flow?

activity 8b
Why must a circuit contact *both ends* of a dry cell to light the bulb?

activity 9
This circuit has a bypass at *a*, another bypass at *b*, and an open switch at *c*. Circle all true sentences about this circuit.

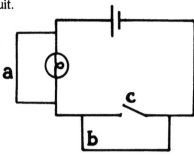

1. The bulb lights if you close *c*.
2. The bulb lights if you remove *a*.
3. The bulb lights if you remove *b*.
4. The bulb lights if you remove *a* and *b*, then close *c*.

activity 10
Redraw this circuit using the correct symbols. Use arrows to show how electrons flow through the circuit.

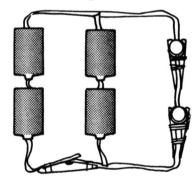

activity 11
Diagram this circuit:

> 1 cell + 1 switch
> 1 cell + 1 switch } in parallel
> 2 bulbs in series }

Use arrows to show how electrons should flow through the wire. You may use your electro-squares, if you wish.

Answers

activity 1

Answers will vary. See notes 1-3 for possible solutions.

activity 2

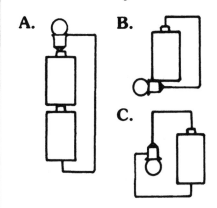

A. **B.** **C.**

activity 3

A. Prediction: no. The bottom contact point on the bulb is not connected to the cell. (The cell shorts out.)

B. Prediction: yes. Both contact points on the cell are connected to both contact points on the bulb.

activity 4

Circle Group A.

Circle Group B.

Circle Group B.

activity 5

a. 4   b. 2   c. 3   d. 1   e. 5

activity 6

Connect a piece of the UFO to a bulb and cell as illustrated. If the bulb shines, it's a conductor. If not, it's an insulator.

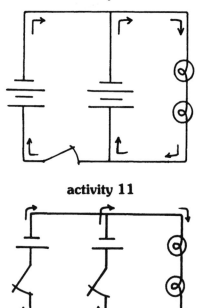

activity 7

a. 1, 5 (Hole 1 is connected with 5 via 4.)

b. 3, 4, 5, or 6. (Holes 1 and 2 are interconnected. If 1 doesn't light with any other hole, neither can 2.)

activity 8a

The positive end of the cell has the bump on it. Since electrons flow from negative to positive, they leave the flat end of the cell, travel through the circuit, and flow back into the bump end of the cell.

activity 8b

To light the bulb, electrons must leave the negative end of the cell and return to the positive end by way of the bulb.

activity 9

Sentences **2** and **4** are true.

activity 10

activity 11

Review / Test Questions, continued

activity 12
Compare each pair of circuits. Circle the one that produces more light.

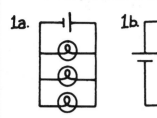

1a. 1b.

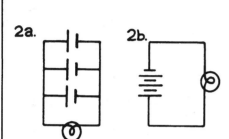

2a. 2b.

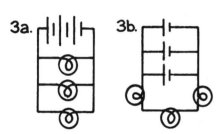

3a. 3b.

In which pair of circuits is the difference in the amount of light the greatest? Why?

activity 13
Add 3 bulbs and 3 switches to this circuit so you can turn off each light independently.

a.

Add 4 bulbs to this circuit so that if you unhook any one bulb, another will go off but the remaining two will stay on.

b.

activity 14
Six copper wires, 3 thin and 3 thick, have different lengths as shown. *Circle* the wire with the greatest resistance. Draw a *box* around the wire with the least resistance.

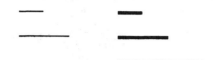

activity 15
To use less copper and thus save money, an electrician installs thinner copper wire in a home than the building code allows. Why is this a dangerous practice?

activity 16
Fill in the blanks with "on" or "off."

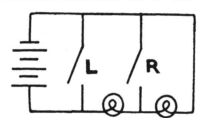

To make *both* lights shine, turn
L _____ and R _____.

To make *one* light shine, turn
L _____ and R _____.

To make *no* lights shine, turn
L _____ and R _____.

activity 17
It's a hot summer evening. All the kitchen lights are on and the air conditioner is humming. You put your dinner in the oven and turn on the temperature control. Suddenly everything goes dark. What's wrong? What should you do?

activity 18
Which steel wool fiber is more likely to get hot enough to pop a balloon, a thick strand or a thin strand? Explain.

activity 19
Which kind of circuit is best to put at the top and bottom of a stairway, A or B? Explain.

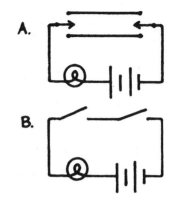

A.

B.

activity 20
Show how electricity flows through each set of bulbs. Indicate which bulbs shine.

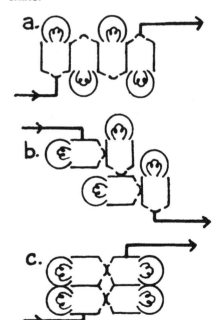

a.

b.

c.

Answers, continued

activity 12

Most light: 1a, 2b, 3a. The third pair has the greatest difference in the amount of light produced. The cells are changed to parallel, producing less light and the bulbs are changed to series, also producing less light.

activity 13

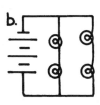

activity 14

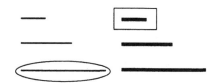

activity 15

Thin wire has higher electrical resistance. It creates a serious safety hazard because the wires could easily overheat and cause a fire.

activity 16

To make both lights shine, turn L <u>off</u> and R <u>off</u>.

To make one light shine, turn L <u>off</u> and R <u>on</u>.

To make no lights shine, turn L <u>on</u> and R <u>off or on</u>.

activity 17

The lights and air conditioner were already drawing maximum power through the line. The oven created an overload and blew a fuse. You should turn off the oven, get a flashlight, and go change the fuse. Then, before heating your dinner, turn off the air conditioner and some of the kitchen lights.

activity 18

A thin strand is more likely to work, because with its higher resistance, it will heat up more quickly to a higher temperature when connected to a cell, and be more likely to burn through the skin of the balloon.

activity 19

Circuit A is better, because you can operate either switch independently. In circuit B, either switch can function only if the other switch remains in an on position.

activity 20

a. The end two only.   b. All   c. All

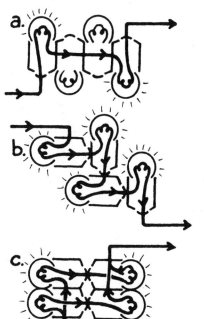

Long-Range Objectives

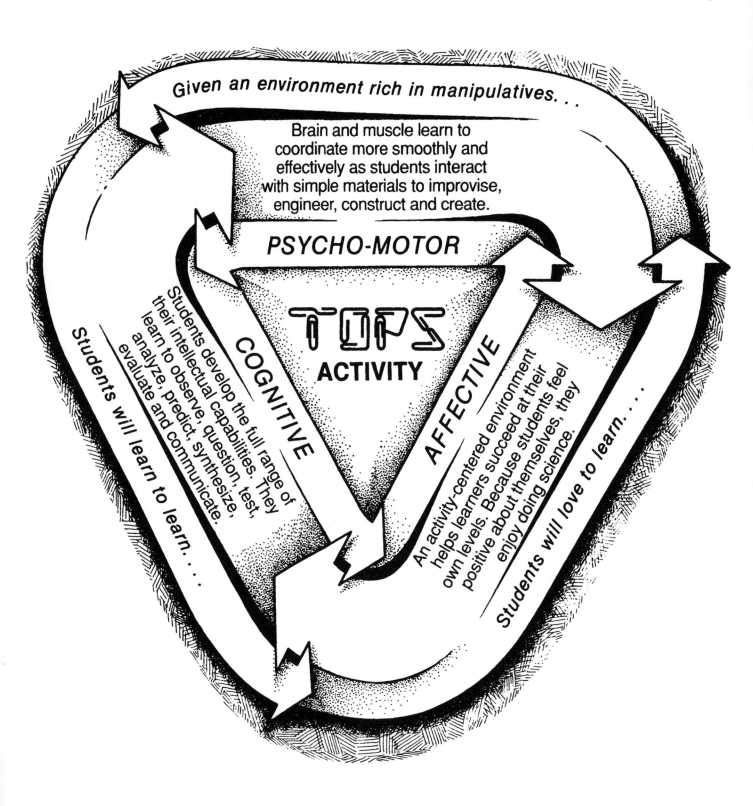

Given an environment rich in manipulatives. . .

Brain and muscle learn to coordinate more smoothly and effectively as students interact with simple materials to improvise, engineer, construct and create.

PSYCHO-MOTOR

TOPS ACTIVITY

COGNITIVE

Students develop the full range of their intellectual capabilities. They learn to observe, question, test, analyze, predict, synthesize, evaluate and communicate.

Students will learn to learn. . . .

AFFECTIVE

An activity-centered environment helps learners succeed at their own levels. Because students feel positive about themselves, they enjoy doing science.

Students will love to learn. . . .

ACTIVITIES
AND
LESSON NOTES
1-20

☞ As you duplicate and distribute these activity pages, **please observe our copyright restrictions** at the front of this book. Our basic rule is: **One book, one teacher.**

☞ TOPS is a small, not-for-profit educational corporation, dedicated to making great science accessible to students everywhere. Our only income is from the sale of these inexpensive modules. If you would like to help spread the word that TOPS is tops, please request multiple copies of our **free TOPS Ideas catalog** to pass on to other educators or student teachers. These offer a variety of sample lessons, plus an order form for your colleagues to purchase their own TOPS modules. Thanks!

IT WORKS!

1 Make some wire from aluminum foil and tape. To do this . . .

30 cm IS A LITTLE LONGER THAN THIS PAGE.

30 cm

. . . stick a piece of tape about 30 cm long to the *dull* side of a strip of aluminum foil.

2 Cut around the edge of the tape to make a foil ribbon.

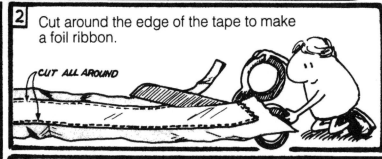

CUT ALL AROUND

3 Fold the ribbon along its length, so the shiny side stays out and the tape is folded inside.

TAPE
SHINY SIDE OF FOIL

4 Crease the fold along the edge of your table.

5

TOPS WIRE FACTORY

REMEMBER HOW YOU MADE THIS RIBBON. YOU'LL NEED MORE LATER.

6 Use this foil ribbon to light a bulb with a dry cell.

??

7 Using pictures like these, draw how you made the bulb light. Also draw a way you tried that didn't work.

DRY CELL

RIBBON

BULB

This works:

This doesn't:

8 Tape your name to your dry cell. You will use it during this whole module on Electricity.

Name

CONSERVE ENERGY: *make your dry cell last!*

TOPS LEARNING SYSTEMS

Objective

To discover by trial and error how to light a bulb with a dry cell and foil ribbon.

Lesson Notes

The foil conducting ribbons used in these lessons are strong, flexible, tangle-proof, and much easier for young children to use than copper wire. Using paper clips with the ribbons, students make secure connections when they build circuits. A completed circuit that is supposed to turn on, turns on – the first time!

1-2. Some students may cut the foil into strips before they apply the tape. This doesn't work. They'll create a mess and have to start over. If you wish to avoid students handling (or mishandling) strips of sticky tape altogether, you can pretape the foil yourself, and have them simply cut out the strips.

Most any kind of tape will work. We recommend that you use ¾ inch masking tape, which folds into convenient ⅜ inch width conducting strips. Alternatively, apply 2 inch packaging tape to the foil and divide it into 3 separate strips.

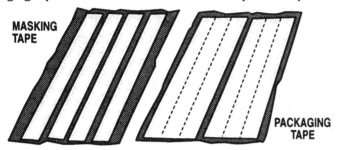

MASKING TAPE

PACKAGING TAPE

3. Aluminum foil conducts electricity, but tape does not. If your students fold the strip with the tape on the outside, they'll end up with a useless strip of non-conducting ribbon. This is less likely to happen if you supply opaque or colored tape; even young children usually guess that the shiny foil side is the "working" part of the ribbon.

But what happens if your students stick clear tape to the foil? They can't see it, so may accidently fold the nonconducting side out. To avoid this, step 1 asks students to apply tape to the *dull* side of the foil, while step 3 directs them to fold the *shiny* side (untaped) to the outside.

6. It may seem surprising that students of high school age, and even adults, don't generally know how to light the bulb. They usually need to do a lot of trial and error investigating. Such exploratory activity is, of course, ideal. Don't be too quick to come to the aid of puzzled students.

Younger students, however, may become frustrated beyond their ability to cope. Help them out by twisting the foil ribbon around the top of the bulb collar (see step 1 of activity 4). With this connection in place, your younger scientists will soon discover the other connections and light the bulb.

Sometimes students complain of being "shocked". Electricity produced by size D cells is totally safe, not strong enough to harm anyone. What they actually feel is the heat generated through the ribbon when it inadvertently connects both poles of the cell. Without a light bulb providing resistance, electron flow is great enough to heat up the ribbon.

Caution students to disconnect these "hot" wires immediately, since they quickly drain the cell of its energy.

ENERGY DRAINING "HOT" WIRE

7. There are many different ways to light a bulb with a cell and wire. Your students will discover many of these variations in activities 1-3.

It's conceptually important to clearly indicate how the contact points on the bulb and cell are interconnected. Watch out for drawings where the bulb looks like an undifferentiated blob, the cell like a box, and the foil ribbon like a railroad track. Refer your students to the schematic drawings provided.

8. This step is optional but recommended. If your students understand that they must continue to use their own cells throughout the entire module, they will more readily conserve energy. Remind them again to immediately disconnect energy-draining hot wires.

Answers

7.

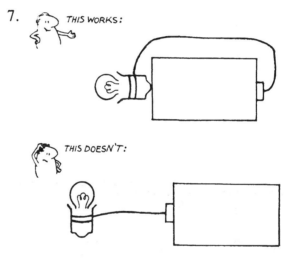

THIS WORKS:

THIS DOESN'T:

Materials

☐ Aluminum foil. Pre-cut rectangles about 30 cm long. See lesson note 1-2.

☐ Masking tape or packaging tape (opaque is desirable. See notes 1-2 and 3.

☐ Scissors.

☐ Size D dry cells (1.5 v), one per student. If fresh, they should last for this entire unit.

☐ Flashlight bulbs, one per student. Use a size designed for 3 dry cells, or 4.5 volts. It may be marked PR 3. It is important to use bulbs from the same manufacturer with the same resistance rating so they will shine equally bright – a requirement for later activities.

TO LIGHT OR NOT TO LIGHT

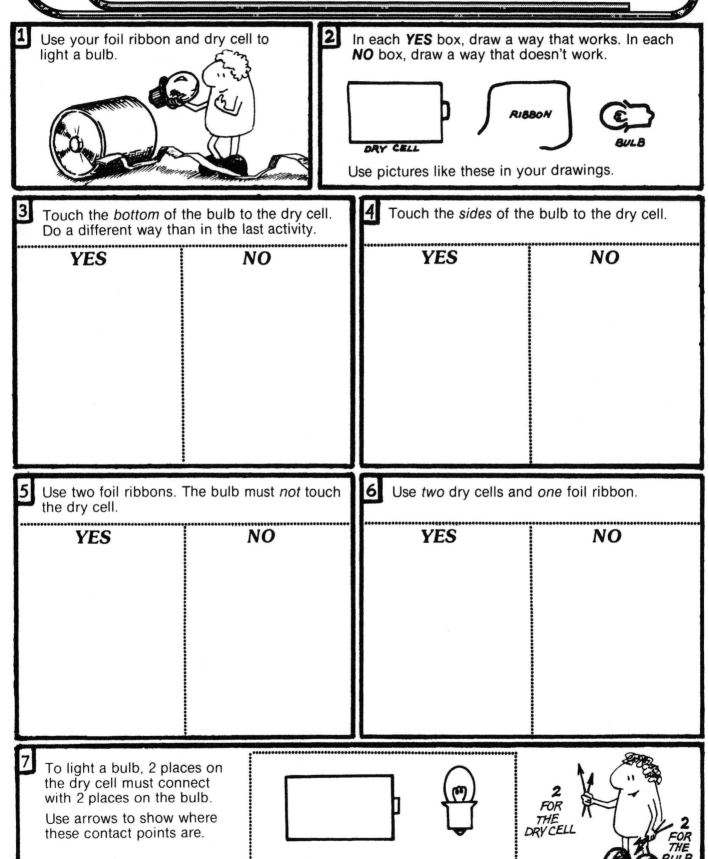

1 Use your foil ribbon and dry cell to light a bulb.

2 In each **YES** box, draw a way that works. In each **NO** box, draw a way that doesn't work.

DRY CELL   RIBBON   BULB

Use pictures like these in your drawings.

3 Touch the *bottom* of the bulb to the dry cell. Do a different way than in the last activity.

YES | NO

4 Touch the *sides* of the bulb to the dry cell.

YES | NO

5 Use two foil ribbons. The bulb must *not* touch the dry cell.

YES | NO

6 Use *two* dry cells and *one* foil ribbon.

YES | NO

7 To light a bulb, 2 places on the dry cell must connect with 2 places on the bulb.

Use arrows to show where these contact points are.

2 FOR THE DRY CELL   2 FOR THE BULB

TOPS LEARNING SYSTEMS

Objective

To further explore by trial and error the different ways to light a bulb with a dry cell and foil ribbon.

Lesson Notes

The activities in this module will provoke a lot of curiosity about electricity. For this reason, a stern warning should be issued to your entire class: **never** fool with a wall plug, socket or other electrical outlet. A severe shock or burn may result. Experiments that go beyond worksheet instructions must first have the teacher's approval.

On the other hand, "what-would-happen-if" questions are at the heart of scientific inquiry, and should be encouraged. Assure the more timid in your class that the electricity produced by a few dry cells is too slight to even tickle. Wires may short-circuit and get warm, but shocks are impossible to come by. Encourage electricity-shy students to set their fears aside, turn their curiosity loose, and enjoy!

1. This activity is a continuation of the first. By trial and error, your students will discover there are many different ways to light a bulb with a cell and ribbon.

At times students may be convinced a bulb ought to light when it doesn't. They may complain that it is burned out or that the cell is dead. This is usually not the problem.

The most common difficulty is simply a short circuit. If, for example, the ribbon inadvertently touches a contact area where it shouldn't, the electrons will by-pass the higher resistance bulb filament completely: no light.

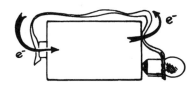

2. Schematic models are again provided to help your students indicate with clarity how the contact points and ribbon interconnect.

3. It is likely that many of your students have already drawn some version of this configuration in activity 1. If so, have them reverse the cell.

4. To find a way that lights, students must work with the bulb and ribbon as separate pieces. If these are joined by twisting at the collar, the dry cell always shorts out when the circuit is completed through the side of the bulb.

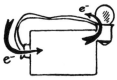 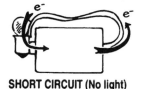

SHORT CIRCUIT (No light)   **SHORT CIRCUIT (No light)**

5. Your students should borrow the second ribbon from a friend, or make another as they did in activity 1. If they make their second ribbon about 20 cm long, it will be the right length to use in step 2 of activity 5.

6. Students who need inspiration here should think about the arrangement of cells in a flashlight. The second cell comes by sharing with a friend.

7. Some may connect the bulb and cell with lines to symbolize wires. Insist that they use arrows. Four arrows emphasize that there are 4 contact points that always need to be connected. This is a crucial concept for success in activity 3.

Answers

3.
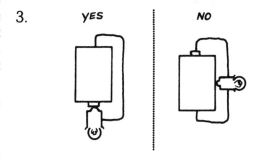

4.
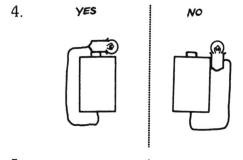

5.
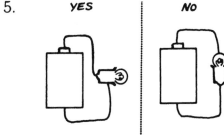

6.
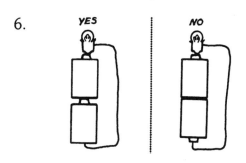

7.
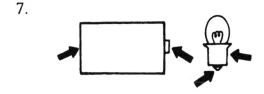

Materials

☐ Foil ribbon from activity 1.

☐ A dry cell and bulb.

LIGHT BULB PREDICTIONS

1

In the table below, guess if the dry cell lights the bulb. Write your *prediction* next to each hook-up.

After you predict, experiment to see if you are right. Write each result in the table.

To make a good prediction, think about HOW MANY contact points must touch to make the bulb light.

2

HOOK-UP	PREDICTION Will it light ?	RESULT Did it light ?		HOOK-UP	PREDICTION Will it light ?	RESULT Did it light ?
A.				D.		
B.				E.		
C.				F.		

If you can't hold all the wires down, ask a friend to help!

3

You are now an expert on how to light a bulb. Write directions for someone who doesn't know how:

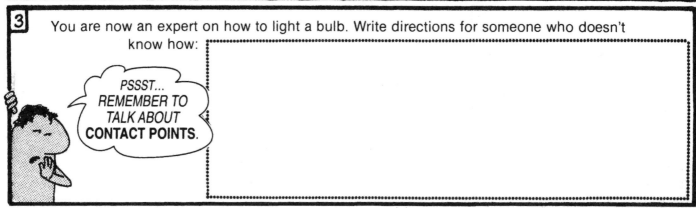

PSSST... REMEMBER TO TALK ABOUT **CONTACT POINTS**.

TOPS LEARNING SYSTEMS

Objective

To use the idea of contact points in predicting whether a bulb will light. To test these predictions by experiment.

Lesson Notes

Notice that the term dry cell (cell for short) is used throughout this module instead of the word battery, a more common usage. A battery is defined as a collection of 2 or more cells wired together. So it's technically incorrect to refer to a single cell as a battery.

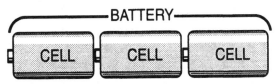

Common English usage, however, ignores this distinction. Even dry cell manufacturers refer to their products as batteries. Your students will, too. That's okay. In time, as they continue to see the terms dry cell and cell in print on their activity sheets, they'll make these words a part of their active vocabularies as well.

1-2. The order of doing things is important. Students must (1) write down their prediction, then (2) do the experiment, and finally (3) write the result. Those who rush ahead and experiment before they predict miss the whole point of this activity: to apply the principle of contact points in a rational, predictive way to bulbs and cells.

2. In parts B and F the bulbs will not light even though electricity flows through the ribbon. Electrons simply pass through the collar of the bulb (but not the filament) on their way from the negative to the positive pole. Because electrons are flowing and the cell is being drained of energy, your students shouldn't leave these short circuits connected too long.

3. Be fussy about this answer. It's time for your students to summarize in their own words what they've learned, and in so doing, solidify the concepts in their own minds. Encourage them to illustrate their answers with diagrams. Those who need extra space should use the back of their worksheet or extra paper.

A CHEAP LIGHT BULB ALTERNATIVE

A 35 to 38 bulb string of steady (not blinking) Christmas tree lights designed for standard 110-volt household outlets may be substituted for 3-volt flashlight bulbs, at a fraction of the cost. These are connected in series, and therefore share voltage: 110 volts ÷ 36 bulbs ≈ 3 volts/bulb. (Each bulb has a shunt, providing an alternate (parallel) pathway should it burn out, allowing the rest of the string to keep shining.)

Cut the string apart, halfway between each light. Strip insulation from the ends of the wire leads, and use masking tape to connect about 6 cm of conducting ribbon to each exposed wire. (To make the connections extra secure, poke a pinhole through the folded ribbon, thread the exposed wire through, and bend it back on itself before taping.)

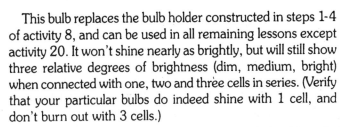

DETAIL:

This bulb replaces the bulb holder constructed in steps 1-4 of activity 8, and can be used in all remaining lessons except activity 20. It won't shine nearly as brightly, but will still show three relative degrees of brightness (dim, medium, bright) when connected with one, two and three cells in series. (Verify that your particular bulbs do indeed shine with 1 cell, and don't burn out with 3 cells.)

Christmas tree lights (with higher resistance) should not be mixed with flashlight bulbs (with lower resistance) in the same circuit. Connected in parallel, they shine with different brightness; in series, the flashlight bulbs won't glow at all.

Answers

2.

	PREDICTION	RESULT
A.	yes	yes
B.	no	no
C.	no	no
D.	yes	yes
E.	yes	yes
F.	no	no

3. Contact points **a** or **b** on the bulb must touch, or be linked by wire to poles **x** or **y** on the cell. The remaining two contact points (one on the bulb and one on the cell) must then be interconnected to complete the circuit and light the bulb.

Materials

☐ Foil ribbon.

☐ A dry cell and bulb.

SERIES MEANS IN A ROW

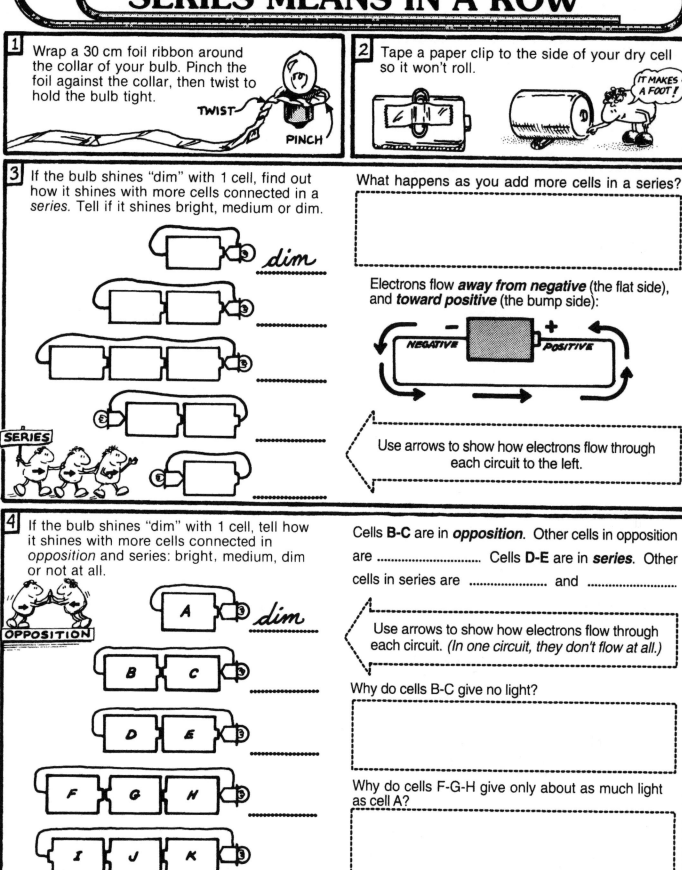

1 Wrap a 30 cm foil ribbon around the collar of your bulb. Pinch the foil against the collar, then twist to hold the bulb tight.

TWIST
PINCH

2 Tape a paper clip to the side of your dry cell so it won't roll.

IT MAKES A FOOT!

3 If the bulb shines "dim" with 1 cell, find out how it shines with more cells connected in a *series*. Tell if it shines bright, medium or dim.

dim

SERIES

What happens as you add more cells in a series?

Electrons flow *away from negative* (the flat side), and *toward positive* (the bump side):

NEGATIVE − + POSITIVE

Use arrows to show how electrons flow through each circuit to the left.

4 If the bulb shines "dim" with 1 cell, tell how it shines with more cells connected in *opposition* and series: bright, medium, dim or not at all.

OPPOSITION

A *dim*

B C

D E

F G H

I J K

Cells **B-C** are in *opposition*. Other cells in opposition are .... Cells **D-E** are in *series*. Other cells in series are .... and ....

Use arrows to show how electrons flow through each circuit. *(In one circuit, they don't flow at all.)*

Why do cells B-C give no light?

Why do cells F-G-H give only about as much light as cell A?

TOPS LEARNING SYSTEMS

Objective

To learn how to connect cells in series and opposition. To understand how this affects bulb brightness.

Lesson Notes

1. Students should use the same aluminum ribbon they made in activity 1. The ribbon's length need only approximate 30 cm, about the length of an actvity sheet.

Be sure the foil is pinched around the collar of the bulb, then twisted firmly to secure a good electrical connection. Joined in this manner, the bulb and ribbon will be used together as one assembly for the remaining activities in this book.

2. The idea here is to keep the dry cell from rolling off the table. When a paper clip is taped perpendicular to the length of the cell, it stabilizes the cell with a kind of foot. Some students may tape the paper clip parallel to the length of the cell. This will not prevent it from rolling.

3. *One* cell is arbitrarily said to shine *dim*. Given this reference, students should use other words like *medium* and *bright*, or *bright* and *brighter* to express how the relative bulb intensities compare.

Physicists arbitrarily think of electric "current" as flowing from positive to negative. Hence **current** is defined as the flow of **positive** charge. This definition is not very useful when applied to the world of cells, bulbs and ribbons. The word *current* is therefore avoided throughout this module and replaced with terms like *electricity* or *flow of electrons*. Because electrons are *negative*, they are shown as **flowing from the negative pole to the positive pole**.

Your students may wonder how electrons flow through the bulb. By now they are aware of the two separate contact areas on the bulb: the outer casing and the end bump. Internal wires connect one end of the light-producing filament to the side casing, and the other end to the end bump. The electrons, of course, can flow in either direction through the bulb, depending on how the cell is connected.

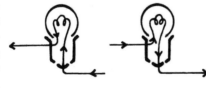

4. Your students may have difficulty determining the direction of electron flow in cells **F-G-H**. Here, the concept is that **G** and **H** in series have more push (voltage) than **F**, in opposition to them both. Electron flow is thus counterclockwise against **F**. To a small degree, **F** is being recharged (see class discussion below).

Cells **F-G-H** actually give a little less light than **A** due to the added internal resistance of 3 cells in the circuit instead of one. An astute observer might write *extra dim* for **F-G-H**.

Discussion

Electrons supplied by chemical reactions within the cell go around the circuit just once. A cell is dead when the chemicals that drive these reactions are nearly used up. The cell can be recharged by using energy to drive the reaction backward, thus using up products to produce new reactants.

Suppose you want to recharge your car battery. Would you connect the charger in series (positive to negative) or opposition (positive to positive)?

Answer: Connect the charger to the battery in opposition; negative to negative and positive to positive. In this way the charger can *pump* electrons into the battery, from positive back to negative.

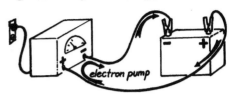

Answers

3.

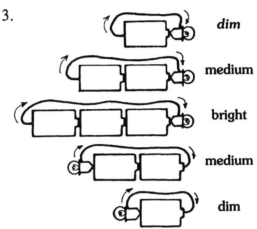

As you add more cells in series, the bulb shines brighter.

4.

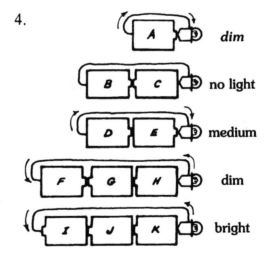

Other cells in opposition: **F - G**

Other cells in series: **G - H** and **I - J - K**

Cells **B-C** give no light because they are in opposition.

Cells **F-G-H** have 2 cells in opposition, leaving only 1 cell to supply electron flow.

Materials

☐ Foil ribbon.

☐ A dry cell and bulb. Students should share dry cells to make at least 3 available per activity group.

☐ Paper clips.

☐ Tape.

PARALLEL MEANS SIDE BY SIDE

1 Start with about 30 cm of foil ribbon attached to your light bulb.

30 cm

2 Make a second foil ribbon about 20 cm long (almost as wide as this paper.)

20 cm

3 If the bulb shines medium with 2 cells in *series*, find out how it shines with cells connected in *parallel*: bright, medium or dim.

medium

PARALLEL
dim

SECOND RIBBON →
dim

dim

Finish each sentence:

When you add more cells *in parallel*, the bulb...
It will remain dim

To make a bulb shine brightest, it is best to connect cells... In series

4 Predict how each bulb shines: bright, medium, dim or not at all. Then experiment to see if you are right.

FIRST PREDICT, THEN EXPERIMENT!

	1. Prediction	2. Result
a. A B	m	
b. C D	dim	
c. E F	not at all	
d. G H I	bright	
e. J K L	dim	

Cells A-B are connected in series .

C-D are connected in parallel .

E-F are in oppisition . G-H-I are in

sereis . K-L are in sereis

while J is in oppistion to them both.

TOPS LEARNING SYSTEMS

Objective

To learn how to connect cells in parallel. To understand how this affects bulb brightness.

Lesson Notes

1. The bulb and 30 cm ribbon come from activity 4.

2. Students may have made this 20 cm ribbon already in activity 2. If not, they should make it now, the same way they made the 30 cm ribbon in activity 1.

3. Bulb brightness is determined by the rate that electrons flow through the filament. Dry cells added in series "push harder" (have more voltage), increasing the electron flow. Dry cells added in parallel increase the available electrons, but since the cells don't push harder, the flow is negligibly increased.

Unless your students have had previous experience with bulbs and cells, this result is quite unexpected. Many will automatically assume, without really looking, that more cells give brighter light and erroneously write *dim, medium, bright* as the cells are added in parallel.

To challenge this preconception, two cells in series at the beginning of this step are arbitrarily said to shine **medium**. Through simple observation your students will have to admit that this medium intensity is much brighter than even 3 cells in parallel. *Dim, dim, dim* is therefore the most appropriate answer.

The best way to compare bulb brightness is to stand three cells on a foil ribbon as shown. Then complete the circuit with a bulb and a second ribbon touching first one cell, then two cells, then all three. Put a penny under the bulb to make a steady glow (better electrical connection).

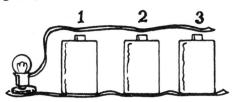

Students will observe that increases in bulb brightness are barely detectable as cells are added in parallel.

4. Look out for students who experiment *before* they predict. They are not exercising their minds, integrating past experience with the problem at hand.

Praise those students who let a *wrong* prediction stand, even after they have come to a correct experimental result! Learning from our mistakes is at the heart of scientific inquiry, and, in fact, those are often the lessons we remember best. Not admitting incorrect predictions (erasing them) is bad science.

The bulbs in boxes **b** and **e** both shine *dim*. There is, however, a slight variation between them. Bulb **e** shines *very dim* because of the internal resistance of 3 cells in the circuit. Bulb **b** shines *slightly* brighter (but still much less than medium) because the circuit has less resistance and more electricity (amperage) supplied by 2 cells, instead of 1.

Answers

3. *medium*
   dim
   dim
   dim

When... more cells in parallel, the bulb <u>remains dim</u>. To make a bulb brightest, connect <u>cells in series</u>.

4. a. medium   medium
   b. dim   dim
   c. not at all   not at all
   d. bright   bright
   e. dim   dim

Cells A-B are connected <u>in series</u>. C-D are connected in <u>parallel</u>. E-F are in <u>opposition</u>. G-H-I are in <u>series</u>. K-L are in <u>series</u> while J is in <u>opposition</u> to them both.

Materials

☐ A dry cell, bulb and foil ribbon. Student should share to make at least 3 dry cells available per activity group.

CONDUCTOR OR INSULATOR?

1 Wind 3 rubber bands around your dry cell as *tightly* as you can.

WIND **ONE** THE **LONG** WAY

THEN WIND **TWO** THE **SHORT** WAY

2 Wrap the free end of your ribbon and bulb 2 times around a penny. Slide the penny between the *flat* end of the dry cell and the rubber band.

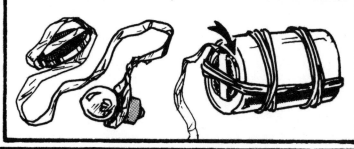

3 Slide the rubber band off the bump on your dry cell. Check to see that the bulb lights.

4 Use your bulb and dry cell to make a list of *CONDUCTORS* and *INSULATORS*. Fill in the table in step 5.

CONDUCTORS LET ELECTRICITY PASS THROUGH

INSULATORS KEEP THE ELECTRICITY FROM PASSING

CONDUCTOR

INSULATOR

5 WASHER? PENNY? GLASS? WOOD?

In what ways are conductors alike?

They are all made from Metal

CONDUCTORS:	INSULATORS:
1. nickle	1. Chulk
2. copper	2. blockbourd
3. lead	3. coms
4. steel	4. skins
5. Iron	5. toothpick
6. tin	6. candie
7. bruss	7. window
8. aluminum	8. eruser

Why do we fold the foil ribbon in half along its length instead of leaving it open?

USE "CONDUCTOR" AND "INSULATOR" IN YOUR ANSWER

TOPS LEARNING SYSTEMS

Objective

To use a bulb and cell to test whether common materials in the classroom are conductors or insulators.

Lesson Notes

1-2. The rubber band, stretched lengthwise, holds the penny in place. It should be wound as tightly as possible. The two rubber bands stretched around the circumference prevent the longer one from slipping off.

The penny presses the ribbon against the dry cell to make an electrically secure connection. (Even if the ribbon doesn't touch the cell directly, this connection still works. As long as the ribbon and cell both touch the penny, the penny serves as an intermediate conducting link.)

4. Some materials – the heating element on an electric stove or the filament in a light bulb, for example – fall *between* these two categories. These might be called poor conductors. Electricity passes through poor conductors, but with enough resistance to give off heat and light.

Conductivity is really a matter of degree. Given high enough voltages, skin is a conductor: an electrician takes every precaution to avoid touching a live wire. But at low voltages, skin is an insulator: your students cannot pass electricity through their fingers to light a bulb – not even through a hangnail!

5. If you allow your students to wander about the room during this step, bulb and cell in hand, they will make wonderful discoveries. They may find, for example, that electricity will travel through a brass door knob or across a filing cabinet. (If it is painted they must touch the bulb and cell to chipped places in the paint, usually along an edge, where the metal is exposed.) Given a long enough ribbon (students can twist or paper-clip them together), they may want to see if electricity will travel the length of your blackboard chalk tray. If it's made of wood, of course, they should be smart enough not to waste time trying. Those with metal dental braces will become instant celebrities, being the only ones in the class with electric teeth.

Younger, less capable students can easily fill out the table by listing the *names of objects*. Challenge older, more able students to list *materials only*, not the object. In the answer key that follows, the "conductors" column on the left side of the table lists only materials (more difficult), while the "insulators" column on the right side lists only objects (easier).

Answers

5.

CONDUCTORS:	INSULATORS:
aluminum	eraser
copper	toothpick
nickel	comb
brass	window
iron	candle
steel	blackboard
lead	chalk
tin	skin

Conductors are all made from metal.

Only the foil side of the ribbon conducts electricity. The tape side is an insulator. To make sure that the ribbon will conduct electricity on both exposed sides, the insulated tape is folded to the middle.

Materials

☐ Rubber bands, at least 3 per student. Size is not important.

☐ A dry cell, bulb and foil ribbon.

☐ A penny.

☐ Test objects. Besides the things in your room, you may want to provide a shoe box filled with a wide assortment of test objects: a washer, button, nail, rubber stopper, piece of steel wool, candle, string and the like.

ELECTRIC PUZZLES

1 Cut some foil the size of an index card.

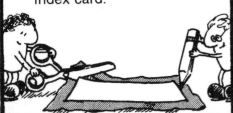

2 Sandwich the foil between 2 index cards.

TAPE THEM TOGETHER

FOIL

3 Fold a *NEW* card in half. Punch 3 holes to make a triangle.

4 Unbend the card. Cover all holes with foil so some are connected and some are not. Fasten with tape.

KEEP SHINY SIDE UP

CONNECTED   UNCONNECTED

5 Paper clip your sandwich over the foil patches to hide them.

← SANDWICH

6 Turn your puzzle over. Number the holes and write your name on it.

7 Unbend a paper clip. Slide it between the bump end of your dry cell and the rubber band.

8 Trade your puzzle for a friend's. Use your bulb and dry cell to find out which holes are connected with foil.

9

Whose puzzle?	Which holes connected?	Results
		☐ Correct ☐ Wrong
		☐ Correct ☐ Wrong
		☐ Correct ☐ Wrong

Find out if you are right. Remove the back sandwich and hold the puzzle up to some light.

1-3 and 4-5!

Then put the puzzle back together so someone else can try it.

TOPS LEARNING SYSTEMS

Objective

To construct a circuit puzzle. To find, by trial and error, those holes that are connected by foil and those that are not.

Lesson Notes

The purpose of this activity will not be completely clear to your students until they reach step 8: It is helpful, therefore, to make a circuit puzzle yourself and show your class its principal parts. First hold up the card with the numbered holes. Show them that only some of these holes are interconnected by foil. Then demonstrate how the foil sandwich prevents puzzle solvers from looking through the index cards to see the solution.

Do not demonstrate how to solve the puzzle with the bulb and dry cell. That is for your students to figure out. Finding which holes are interconnected by foil is, of course, simply a matter of finding the combinations that light the bulb. Your students must exercise care, however, to conduct a methodical search: there are 15 different double combinations to consider.

2. The foil layer makes the sandwich completely opaque. Even when you hold it to the light, you can't see through.

3. Despite explicit directions to the contrary, there are always a few who will fold and punch the foil sandwich. Back to step 1 for them.

4. If different sides of the foil are placed face down, some holes will appear shiny and others dull. Holes that look identical are more likely (though not necessarily) interconnected. If the same side of the foil shows through all the holes, this contextual clue will be eliminated.

7. Activity 6, step 2 shows how the foil ribbon is wound around a penny and attached to the flat end of the cell. This is out of view in this illustration.

8. If your class is organized on an individualized basis, and there are not enough other students who have reached this step, you might provide some of your own puzzles to trade. Here are a few suggestions.

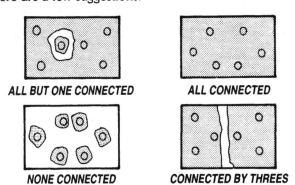

ALL BUT ONE CONNECTED   **ALL CONNECTED**

NONE CONNECTED   **CONNECTED BY THREES**

9. There is room in the table for each student to try solving the puzzles of at least three friends. Some (especially younger students) may be eager to solve additional puzzles. They can extend this table to the back of their worksheets.

Answers

9. varied answers

Materials

☐ Index cards, preferably 3x5 inches.

☐ Aluminum foil.

☐ Pair of scissors.

☐ A paper punch.

☐ Tape.

☐ Paper clips.

☐ A dry cell, bulb and foil ribbon. These may already be assembled from activity 6.

BUILD A CIRCUIT

1 Get 2 clothespin halves.

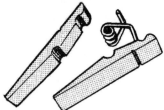

Clamp these around your bulb and foil ribbon.

2 Place a paper clip so it touches the bulb and rests on the collar.

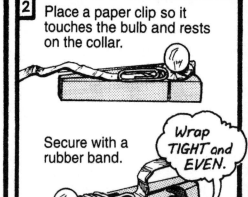

Secure with a rubber band.

Wrap TIGHT and EVEN.

3 Put a third half of a clothespin under the first two.

Wrap another rubber band around the end.

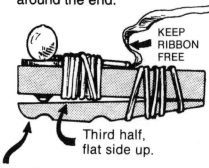

KEEP RIBBON FREE

Third half, flat side up.

Thick end under bulb.

4 Slip a coin between the bulb and clothespin. Slide a 20 cm foil ribbon under this coin.

5 Make 2 foil ribbons that are each 12 cm long.

12 cm is as long as 2 dry cells.

← 12 cm →

6 Wrap each ribbon 2 times around a coin, and slide it under the rubber band at each end of your dry cell.

7 Hook your bulb holder and dry cell together with paper clips.

I made a CIRCUIT.

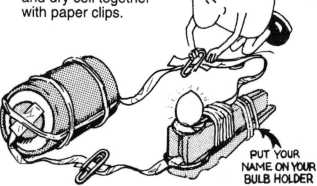

PUT YOUR NAME ON YOUR BULB HOLDER

What path do electrons take as they move around your circuit?

TOPS LEARNING SYSTEMS

Objective

To make a bulb holder, and complete construction of a dry cell holder for use in future activities.

Lesson Notes

Construct a bulb and dry cell holder yourself, before your students try. This will familiarize you with the directions, and provide a model for your students to follow.

1. These clothespin halves must be pulled apart just once. (You can throw the spring away.) Thereafter, you can recycle the halves to use again each time you teach this series of activities.

The foil ribbon remains crimped around the collar of the bulb as in activities 4-7. Make sure your students have pinched this foil to the collar as tightly as possible (fingernails work well) so no excess foil sticks up. This insures that the paper clip, bulb and clothespin fit snugly together in step 2.

2. The paper clip wrapped in the rubber band *must* overlap the collar of the bulb. This presses the bulb firmly against the top of the clothespin, so it cannot slide up when pushed from below by another clothespin half in step 3.

If your bulbs have screw-in threads without collars, omit the paper clip entirely. Instead, tightly wrap the rubber band directly around the clothespin and foil ribbon, as near to the bulb as possible. If wrapped tightly, threads in the bulb's neck should bite into the soft wood of the clothespin, holding it firmly in place.

3. Wrap the third clothespin half with the other two, so the tapered ends of all three press together, as illustrated. Wind this rubber band with less tension — just tight enough to hold the bottom clothespin half firmly yet gently against the bottom of the bulb. Lift up the foil ribbon to keep it entirely free of this second rubber band.

4. The penny is important. It forms a broad, flat electrical link between the contact point of the bulb above and the foil ribbon below. Without the penny in place, the bulb may flicker due to a poor electrical connection.

5-6. Students should make the ribbons as they did in the first activity.

After the ribbon is wrapped two full turns around the penny, there won't be much left to stick out beyond the ends of the cell. This is ideal. Longer ends tend to inadvertently touch and short out the cell, rapidly draining its energy.

A good way to ensure long cell life is to trim each ribbon a little too short to touch the other.

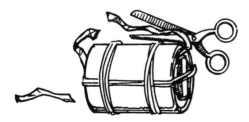

7. Remind your students to write their names on their bulb holders. They will use them many times in later activities.

Inspect each student's bulb and cell holder while they are paper-clipped into a complete circuit. The bulb should shine steadily without flickering, even if the circuit is moved about the table.

TROUBLE SHOOTING

a. Recheck your connections. Do the paper clips press both ribbons firmly together?

b. Test each component in the circuit with other equipment that you know is working.

c. Examine your foil ribbons. Is the tape always folded to the inside? Has any ribbon been stretched so hard that the foil has split?

d. Shine up dull contact points by rubbing with steel wool. Over time, aluminum and copper surfaces may tarnish, building up layers of insulating oxides.

Answers

7. Electrons flow from the flat end (negative pole) of the cell, through the ribbon, then through the penny, then through the bulb, and finally back through the other ribbon to the bump end (positive pole) of the cell.

Materials

☐ Paper clips.

☐ A dry cell and bulb.

☐ Clothespins.

☐ Rubber bands.

☐ Foil ribbon. Students will need to make additional ribbons as they did in activity 1. Provide aluminum foil, tape and scissors.

☐ A penny or other coin.

ELECTRIC BY-PASS

1 Make a *switch:* Tape foil ribbons about 12 cm long (as long as 2 dry cells) to each end of a clothespin half.

OVERLAP THE ENDS

TAPE HERE   TAPE HERE

2 Bend the top ribbon so it touches the bottom ribbon *only* when you push down.

PUT YOUR NAME ON *YOUR* SWITCH

3 Build a circuit: Connect your bulb, dry cell and switch with paper clips.

Tell how your switch is able to turn the light on and off:

4 You can keep your switch turned on with another clothespin.

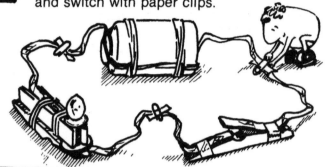

You can *by-pass* the bulb at **A,** the switch at **B,** or the ribbon at **C** by putting a foil bypass across them like this:

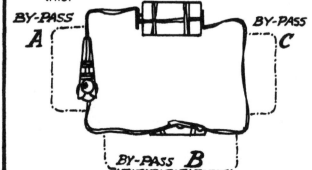

BY-PASS **A**   BY-PASS **C**

BY-PASS **B**

	WHAT HAPPENS?	WHY?
While the switch is ON, put a ribbon across **A**		
While the switch is OFF, put a ribbon across **B**		
While the switch is ON, put a ribbon across **C**		

TOPS LEARNING SYSTEMS

Objective

To make a switch and integrate it into a simple circuit. To study how alternate pathways around the bulb and switch affect the circuit.

Lesson Notes

1. The ribbons must not be taped where they overlap, since this would insulate the contact points and render the switch useless. Your students are less likely to make this mistake if you provide easy-to-see masking tape rather than clear tape.

2. Remind your students to write their names on their switches. They will use them again in later activities.

3. This question can also be answered in terms of contact points, a concept developed in activities 1-3. When the switch is opened, the circuit is broken: the contact point on the end of the bulb is no longer connected to a pole on the cell.

4. If extra ribbons are not available, your students will need to make them as they did in activity 1.

By-pass A shorts out the bulb, while by-pass B shorts out the switch. This concept of short circuits is treated more fully in activity 16.

Extension

CAR WIRING

Here's a mystery for your class to solve. Wires lead out from the positive terminal of your car battery to all the lights in and on your car. But no wires lead from these same lights back to the negative terminal.

You can demonstrate this phenomenon under the hood of a car in your school's parking lot. You'll need an auto light bulb designed for 12 volts, plus a length of *insulated* wire. Also take a flashlight if it's a cloudy day.

Strip the ends of the wire and wrap one end around the collar of the bulb. You can light it by touching the bottom of the bulb to any metal part of the car frame not insulated from the engine chassis or painted, and the free end of the wire to the positive terminal of the car battery. (Caution: Use *insulated* wire only. If you inadvertently touch a *bare* wire connected to the positive terminal to the car frame, sparks will fly in a spectacular short circuit.)

How do electrons get to the bulb? The mystery clears when you point out that electrons travel from the negative terminal into the car frame! (Point out the grounding wire that connects the battery to the engine chassis.) The bulb lights because the car frame completes the circuit: electrons run from the negative terminal, through the grounding wire to the frame; through the frame to the bulb, then back through the connecting wire to the positive terminal of the battery.

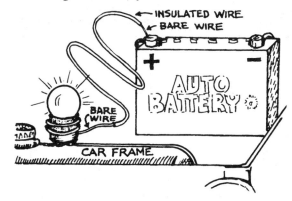

Answers

3. The bulb lights only as electrons flow through it. When the switch is turned off, this flow is blocked; electrons can't cross the gap.

4. A. The bulb goes out. — Electrons travel through the by-pass instead of the bulb.

   B. The bulb turns on. — Electrons can now flow around the gap created by the switch.

   C. The bulb continues to shine. — Nothing is by-passed except some wire. Electrons can travel both routes to complete the circuit.

Materials

☐ A clothespin half to make the switch. The bulb holder in activity 8 required three clothespin halves. There should be one left over.

☐ Tape. See note 1 above.

☐ The dry cell holder, bulb holder and foil ribbon from activity 8.

☐ Paper clips.

CIRCUIT SYMBOLS

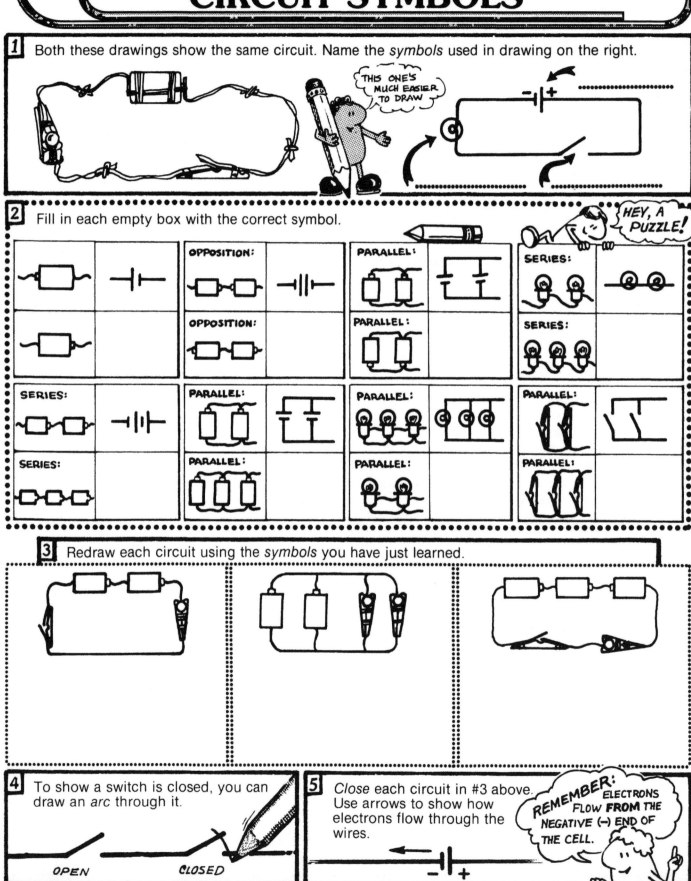

1 Both these drawings show the same circuit. Name the *symbols* used in drawing on the right.

THIS ONE'S MUCH EASIER TO DRAW

2 Fill in each empty box with the correct symbol.

HEY, A PUZZLE!

OPPOSITION:

OPPOSITION:

PARALLEL:

PARALLEL:

SERIES:

SERIES:

SERIES:

SERIES:

PARALLEL:

PARALLEL:

PARALLEL:

PARALLEL:

PARALLEL:

PARALLEL:

3 Redraw each circuit using the *symbols* you have just learned.

4 To show a switch is closed, you can draw an *arc* through it.

OPEN   CLOSED

5 *Close* each circuit in #3 above. Use arrows to show how electrons flow through the wires.

REMEMBER: ELECTRONS FLOW **FROM THE** NEGATIVE (—) END OF THE CELL.

– | | +

TOPS LEARNING SYSTEMS

Objective

To learn to draw simple circuit diagrams using accepted symbols. To predict how electrons should flow through the wires.

Lesson Notes

Scientists and engineers draw circuit diagrams to show how bulbs and switches and such are interconnected. Their squiggles and lines strongly suggest the form of the part they wish to represent. Students will quickly master these symbols and take great delight in expressing the interconnectedness of things using their strange new language.

This entire activity is a written exercise. Beyond pencil and paper, no special materials are required. Be sure your students do use a pencil with a good eraser, not a pen. As with any new language, they are bound to make errors that will require corrections. Insist that these corrections are made neatly and that careless exercises are done over. An incomprehensible tangle of lines and smudges should not qualify as an acceptable diagram.

2-3. The symbol for cells in a battery seems most difficult to learn. Common errors include the following:

Positive and negative poles can't be distinguished.

Connecting wires between cells in series are drawn in.

Long gap, as if the body of the cell should fit between.

A good way to remember that the longer line represents the positive pole is to recall the internal structure of a cell. In a zinc-carbon dry cell, the bump at the positive end caps a long carbon rod that extends down through the black electrolyte. Think of the longer line as representing this carbon rod. Since it receives electrons from the zinc can, it must be positive.

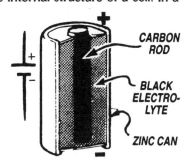

CARBON ROD

BLACK ELECTRO-LYTE

ZINC CAN

Answers

1. dry cell, bulb, switch

2.

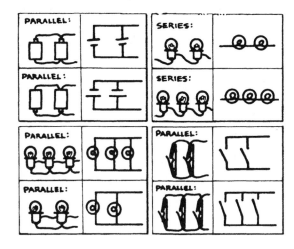

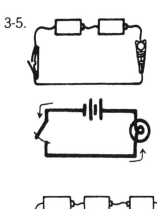

3-5.

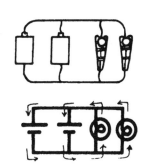

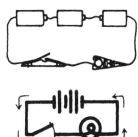

(Cells in opposition can't push electrons both ways at the same time. In this case they are pushed counterclockwise by the higher potential of the two cells in series.)

Materials

None

ELECTRO-SQUARES

1 Find the other paper called ELECTRO-SQUARES. Cut out each square along the *dashed* lines.

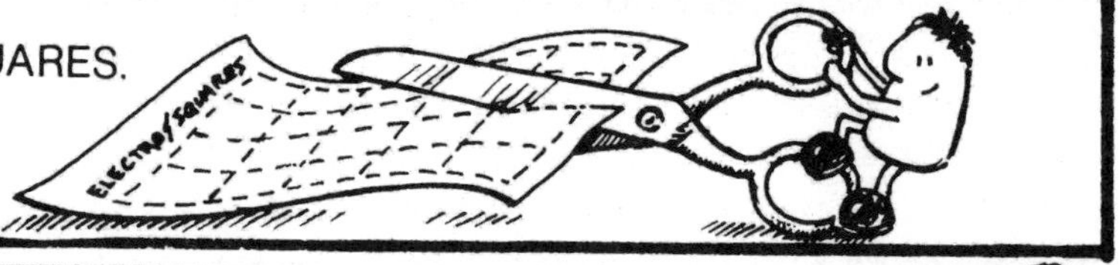

2 For each box below:

MAP the circuit with your ELECTRO-SQUARES.

DRAW the circuit in the box.

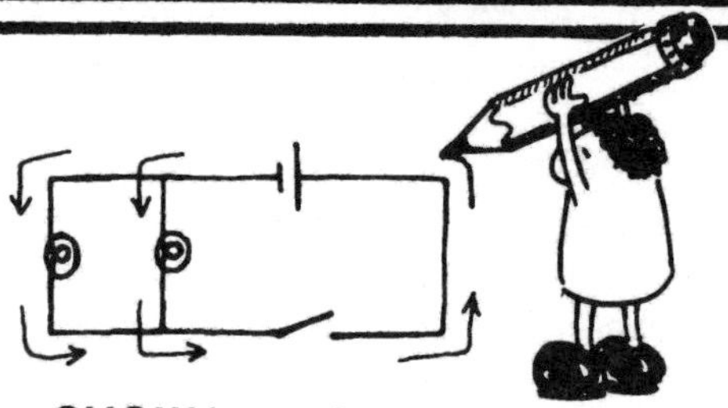

SHOW how electrons **FLOW** through the circuit.

a. 1 cell + 1 bulb + 1 switch:

☐ *Map* ☐ *Draw* ☐ *Show flow*

b. 3 cells in series with 1 bulb + 1 switch

☐ *Map* ☐ *Draw* ☐ *Show flow*

c. 3 bulbs in parallel with 1 cell + 1 switch

☐ *Map* ☐ *Draw* ☐ *Show flow*

d. 2 cells in series + 2 bulbs in series + 1 switch:

☐ *Map* ☐ *Draw* ☐ *Show flow*

e. 2 cells in parallel + 2 bulbs in parallel

☐ *Map* ☐ *Draw* ☐ *Show flow*

f. 1 bulb + 1 switch } in parallel with
1 bulb + 1 switch } 2 cells in series

☐ *Map* ☐ *Draw* ☐ *Show flow*

g. 1 cell
1 bulb } in parallel
1 switch

☐ *Map* ☐ *Draw* ☐ *Show flow*

h. 2 cells in opposition }
1 cell + 1 bulb } in parallel
1 switch + 1 bulb }
1 switch + 1 bulb }

☐ *Map* ☐ *Draw* ☐ *Show flow*

3 Paper clip your ELECTRO-SQUARES into one pile. You'll need to use them again.

TOPS LEARNING SYSTEMS

Objective

To practice mapping and drawing more complicated circuit diagrams. To predict how electrons should flow through these circuits.

Lesson Notes

1. These *electro-squares* are used in activities 11, 12, and sometimes 13. Make extra copies for students who cut them wrong or lose some of their squares and need replacements.

2. Check boxes are provided for each circuit so that students will complete all three steps before proceeding to the next problem. These squares can be self-checked by students, or teacher-checked if you wish to monitor student progress more closely.

Again, make sure that students use pencils so they can erase errors. And insist on neatness: the diagrams should be

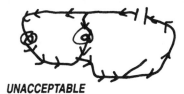

UNACCEPTABLE

rectangular in shape with sharp corners. Electron flow should be indicated by just a *few* arrows drawn to the side of the wires, not on the wires.

To familiarize students with the symbol for a dry cell, the cell shape is lightly shaded on the electro-squares. When students diagram each circuit after mapping it, they should copy *only the symbol*, not the dry cell image itself.

The diagrams students draw, of course, may have the cells, bulbs and switches drawn in different order. But the overall circuit paths, dictated by series or parallel construction, must not vary.

The electro-squares serve as an intellectual crutch, helping students piece together circuits they might not otherwise be able to draw. As they progress through this lesson, some may begin to diagram circuits directly, without mapping first. This is great: concrete manipulation of the electro-squares has enabled these students to advance to a higher level of mental abstraction. The electro-squares are no longer needed!

2a. To avoid excessive wordiness in these problems, terms like *1 cell + 1 bulb + 1 switch* are assumed to mean *in series* unless otherwise stated.

2b. Here, if students try to line up the cells, bulb and switch all on the same side, they will run out of squares. Trial and error will help them learn to use their resources wisely.

2c. The 3 bulbs are to be connected *in parallel*, not series. Students who have trouble here should look at the illustration just above box C for parallel inspiration. You might also suggest that they use all four "T"-squares.

2g. Here the electrons flow through the bulb only when the switch is *open*. When the switch is closed, the bulb is bypassed. Your students will build this backward-switch circuit in activity 16.

2h. Some students may show electrons flowing out both ends of the cells in opposition. This is incorrect. There **must** be a balance between the number of electrons leaving and entering a cell. The positive pole of one cell can not accept the electrons produced by the negative poles of three other cells without creating a charge imbalance.

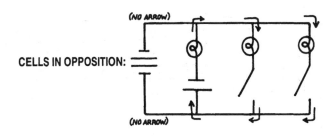

CELLS IN OPPOSITION:

Answers

2. a. 1 cell + 1 bulb + 1 switch:

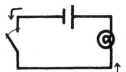

b. 3 cells in series with 1 bulb + 1 switch:

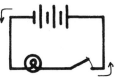

c. 3 bulbs in parallel with 1 cell + 1 switch:

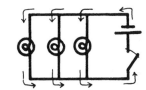

d. 2 cells in series +2 bulbs in series + 1 switch:

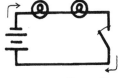

e. 2 cells in parallel + 2 bulbs in parallel:

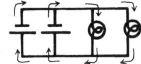

f. 1 bulb + 1 switch } in parallel
1 bulb + 1 switch } with 2 cells in series:

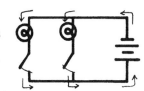

g. 1 cell
1 bulb } in parallel:
1 switch

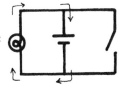

h. 2 cells in oppos.
1 cell + 1 bulb
1 switch + 1 bulb } in parallel:
1 switch + 1 bulb

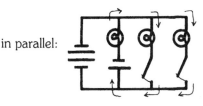

Materials

☐ Duplicated electro-square cut-out sheets.

☐ Pair of scissors.

MAP IT – DRAW IT – BUILD IT

1 In parts 2 and 3 below . . .

MAP each circuit
with your ELECTRO-SQUARES . . .   **DRAW** each circuit . . .   **BUILD** each circuit.
COMPARE how bright the bulbs shine.

2 *2 cells in series* with 1 bulb + 1 switch in series.   *2 cells in parallel* with 1 bulb + 1 switch in series.

☐ Map it... ☐ Draw it... ☐ Build it!   ☐ Map it... ☐ Draw it... ☐ Build it!

Two CELLS in series make the bulb shine .... than 2 cells in parallel.

3 *2 bulbs in series* with 2 cells + 1 switch in series.   *2 bulbs in parallel* with 2 cells + 1 switch in series.

☐ Map it... ☐ Draw it... ☐ Build it!   ☐ Map it... ☐ Draw it... ☐ Build it!

Two BULBS in series shine .... than two bulbs in parallel.

TOPS LEARNING SYSTEMS

Objective

To learn how to connect dry cells and bulbs to acheive maximum brightness.

Lesson Notes

1. You might wish to make mapping with electro-squares optional, depending on the ability of your students. However, those who draw the circuit incorrectly should be encouraged to map their way to the correct answer.

Even reliable TOPS circuits may refuse to work now and then. If so, consult the trouble-shooting checklist in the teaching notes of activity 13.

2. Three check squares in each box emphasize that there are three distinct tasks to be completed for each circuit. If you or a respected lab aid check off these squares for your students, they will probably work harder and faster! Checks are motivational, much like a word of praise or pat on the back. More capable students, of course, can easily monitor their own progress with self-checks.

When comparing series and parallel connections, it is important to control variables by using the **same** two dry cells.

The same **cells** in series make the bulb shine brighter because you increase the voltage – the electromotive force carried by each electron. With the cells in parallel, you increase only the amperage – the amount of electron flow. This is analogous to water over a dam. To make a water wheel below spin more rapidly, you would raise the dam (increase the number of volts) rather than add more water (increase the number of amps).

3. For purposes of comparison, both bulbs in the circuit should have the same resistance and be from the same manufacturer. This insures that the bulbs will shine with comparable intensity when connected to the cells.

Again, the same two cells and bulbs should be used for both circuits. With these variables controlled, the **bulbs** in series shine much dimmer than the bulbs in parallel. When connected in series, each bulb must share the voltage delivered by the cells. When connected in parallel, each bulb has an independent path to the cells, so that each one receives full voltage. Using the same analogy, water wheels spin faster in the falls when placed side-by-side (in parallel). If the wheels are linked together so that one is directly over the other (in series), they will both use the same energy from the falling water, and thus spin more slowly.

Answers

2. **2 _cells_ in series** with 1 bulb + 1 switch in series:

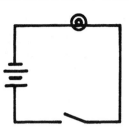

2 _cells_ in parallel with 1 bulb + 1 switch in series:

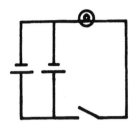

Two CELLS in series shine **brighter** than two bulbs in parallel.

3. **2 _bulbs_ in series** with 2 cells + 1 switch in series:

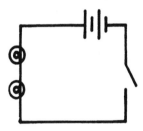

2 _bulbs_ in parallel with 2 cells + 1 switch in series:

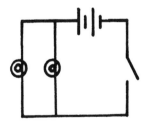

Two bulbs in series shine **dimmer** than two bulbs in parallel.

Materials

☐ Electro-squares.

☐ Circuit components: dry cell holder, bulb holder, switch, foil ribbon and paper clips. Students should share equipment to make at least 2 cells and 2 bulbs available per activity group.

SERIES OR PARALLEL?

1 Draw this circuit, then build it:

1 bulb + 1 switch ⎫
1 bulb + 1 switch ⎭ in *series* with 2 cells in series

Do the bulbs shine brightly or dimly?..

Will one switch turn on one light? Why?

If a bulb burns out will the other still shine? Why?

2 Draw this circuit, then build it:

1 bulb + 1 switch ⎫
1 bulb + 1 switch ⎭ in *parallel* with 2 cells in series

Do the bulbs shine brightly or dimly?..

Will one switch turn on one light? Why?

If a bulb burns out will the other still shine? Why?

3 Suppose you're an electrician. Would you wire a house in a series or parallel? Give 2 reasons.

4 Suppose you make Christmas tree lights.

Is it easier and cheaper to wire them in series or parallel?

Which way of wiring makes the best product?

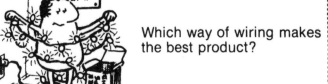

TOPS LEARNING SYSTEMS

Objective

To understand why electricians wire buildings in parallel rather than in series.

Lesson Notes

1-2. This activity demonstrates some of the advantages that parallel circuits have over series circuits. The questions are easiest to answer if your students leave their circuits assembled to use as a point of reference. Those who immediately pull them apart are most likely to miss the important conceptual connections.

Even under the best of circumstances, things can go wrong. If students can't make a particular circuit work, refer them to the following checklist. Let your students solve their problems for themselves, if at all possible.

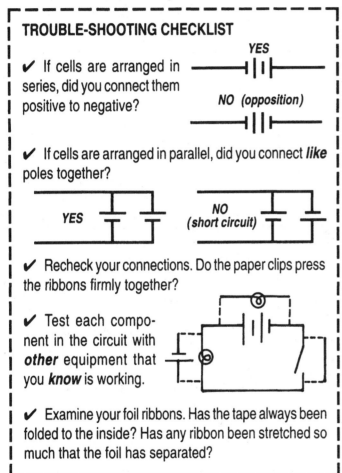

TROUBLE-SHOOTING CHECKLIST

✔ If cells are arranged in series, did you connect them positive to negative?

YES

NO (opposition)

✔ If cells are arranged in parallel, did you connect *like* poles together?

YES

NO (short circuit)

✔ Recheck your connections. Do the paper clips press the ribbons firmly together?

✔ Test each component in the circuit with *other* equipment that you *know* is working.

✔ Examine your foil ribbons. Has the tape always been folded to the inside? Has any ribbon been stretched so much that the foil has separated?

1. This circuit is relatively easy to build. It hooks together in a large, continuous circle.

2. Because of its parallel construction, this circuit is harder to build. Students who experience difficulty here should map it first with their electro-squares.

4. Your students may wonder how Christmas tree lights are wired together into what appears to be a single wire. An electric cord that is wired in parallel has two insulated wires

enclosed in one single sheath. A blackboard sketch might help you illustrate this point:

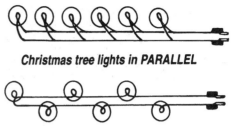

Christmas tree lights in PARALLEL

Christmas tree lights in SERIES

Be sure to make this point again: even though there are similarities, AC/DC 110 volt light bulbs and wall sockets are not to be equated with bulbs and dry cells. Messing around with household electricity is foolish and dangerous.

Answers

1.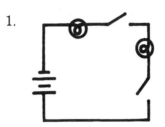

The bulbs shine <u>dimly</u>.

No. One switch in series can only turn on *both* lights as long as the other switch stays closed.

No. The burned-out bulb breaks the circuit for the other bulb, too.

2.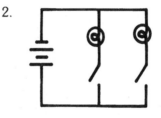

The bulbs shine <u>brightly</u>.

Yes. Each bulb has an independent path to the cells through its own switch.

Yes, because the other bulb still has an unbroken, independent path to the cells.

3. You should wire a house in parallel:
   Each appliance and light receives full power;
   Each can be turned on/off without affecting the others;
   If one burns out, the others stay on.

4. *Series* wiring is easier and cheaper, but *parallel* wiring would make a better product.

Materials

☐ Electro-squares (optional).

☐ Circuit components: dry cell holder, bulb holder, switch, foil ribbon and paper clips. Students should share equipment to make at least 2 cells, 2 bulbs and 2 switches available per activity group.

RESISTANCE IN A WIRE

1 Divide an index card in two.

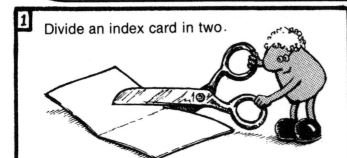

Pull off a single strand of steel wood and tape each end to one of the halves:

Unbend a paper clip and tape it to the other half:

2 Unbend another paper clip. Slide it under the penny on your bulb holder, replacing the foil ribbon.

3 Connect your bulb to 2 cells in series. Attach a foil ribbon and paper clip to the other end.

4 Pass electrons through the *THICK* paper clip and the *THIN* steel wool. Compare how bright the bulb shines.

THICK WIRE   **THIN WIRE**

FILL IN EACH SPACE WITH THESE WORDS

Shines brighter: ....

Shines dimmer: ....

Holds back the flow of electrons: ....

Allows electrons to flow: ....

Has low resistance: ....

Has high resistance: ....

5 Pass electrons through the steel wool for a *LONG* distance and a *SHORT* distance. Compare how bright the bulb shines.

LONG WIRE   **SHORT WIRE**

Shines brighter: ....

Shines dimmer: ....

Holds back the flow of electrons: ....

Allows electrons to flow: ....

Has low resistance: ....

Has high resistance: ....

6 Resistance in a wire *INCREASES* as . . .

Resistance in a wire *DECREASES* as . . .

(name 2 ways)

(name 2 ways)

TOPS LEARNING SYSTEMS

Objective

To understand how length and diameter affect electrical resistance in wire.

Lesson Notes

1. A *single* strand of fine grade steel wool has a diameter comparable to a human hair. They can look so much alike, in fact, that students have sometimes taped hair to the card by mistake, and then wondered why the bulb wouldn't light.

Students should exercise care in pulling strands from the ball of steel wool, handling it lightly with the tips of their fingers. Otherwise they are subject to painful steel slivers. Just in case, be sure your medical kit contains tweezers.

2. In this activity and the next, a straightened paper clip replaces the foil ribbon that is usually pressed under the penny on the bulb holder.

3. The bulb should shine brightly when the two paper clip probes are touched together. If not, consult the trouble-shooting checklist in the teaching notes of activity 13. The most common problem results from connecting the cells in opposition rather than series.

4. If students place the probes too close together on the steel strand, the strand may spark and melt into two pieces. This little surprise, generating considerable interest, is fully exploited in activity 15. For now, tell your students to re-string the half index card with another strand and try again. This time they should begin with the probes farther apart, then slowly slide them nearer. The strand is in danger of melting only when the electricity passes through a very short segment between very close probes.

4-6. The completion questions in steps 4 and 5 are designed to lead your students, in step 6, to logical conclusions about resistance based on their observations of bulb brightness. Students who experience difficulty here need more opportunity to make *resistance* a part of their working vocabularies. You can help by doing the following extension with your class, followed by a general discussion of resistance.

Extension

Electrical resistance in wire is analogous to forcing air through a tube. It is easier to blow air through a wider tube than through a narrower tube; through a shorter tube than through a longer tube.

a. Distribute 1 straw and 1 piece of scratch paper to each student. Make scissors and tape accessible to all.

b. Cut the straw into one long piece and one very short piece.

c. Roll, cut and tape the paper to make a wide tube equal in length to the longer straw.

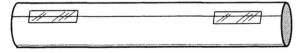

d. Now blow air through each tube as hard as you can. Which offers more resistance, the wider long paper tube or the narrower long straw tube? The longer straw or the shorter straw?

Answers

4. Shines brighter: <u>thick wire</u>
   Shines dimmer: <u>thin wire</u>
   Holds back flow of electrons: <u>thin wire</u>
   Allows electrons to flow: <u>thick wire</u>
   Has low resistance: <u>thick wire</u>
   Has high resistance: <u>thin wire</u>

5. Shines brighter: <u>short wire</u>
   Shines dimmer: <u>long wire</u>
   Holds back flow of electrons: <u>long wire</u>
   Allows electrons to flow: <u>short wire</u>
   Has low resistance: <u>short wire</u>
   Has high resistance: <u>long wire</u>

6. Resistance in a wire INCREASES as:
   thickness decreases and/or length increases.

   Resistance in a wire DECREASES as:
   thickness increases and/or length decreases.

Materials

☐ An index card, preferably 3x5 inches.

☐ Pair of scissors.

☐ Steel wool. Usually available in hardware and drug stores, sometimes supermarkets. Avoid presoaped pads or the coarser industrial grades. Select a fine grade of wool about as thick as human hair.

☐ Tape.

☐ Circuit components: dry cell holder, bulb holder, foil ribbon and paper clips. Students should share to make at least 2 cells available per activity group.

A FLASHY EXPERIMENT

1 Connect your bulb to dry cells in series. Attach a foil ribbon and straight pin to the other end.

PINCH

2 Gently pull on a piece of steel wool so iron fibers fall on an index card.

BE CAREFUL NOT TO GET SLIVERS IN YOUR FINGERS

3 Pass electrons through the tiny steel wool fibers on your card. Observe what happens to the steel wool *and* the light bulb.

A. What happens when electrons pass through very *THIN* strands?

B. What happens when electrons pass through *THICK* strands or clumps?

C. How does "resistance" explain the difference between *a* and *b*?

4 Order these materials by how much resistance they have:

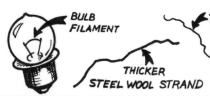

BULB FILAMENT

THINNER STEEL WOOL STRAND

THICKER STEEL WOOL STRAND

FOIL RIBBON

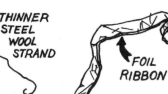

1 .... HIGHEST

2 ....

3 ....

4 .... LOWEST

TOPS LEARNING SYSTEMS

Objective

To watch a miniature fireworks show! To appreciate that high electrical resistance can create heat and light.

Lesson Notes

1. The bulb holder has a straightened paper clip probe, instead of a foil ribbon, projecting from beneath the penny. Students made this exchange in step 2 of activity 14.

For this experiment to be really flashy, the bulb should shine brightly when the pin and paper clip probes are touched together. If fresh cells are used, two in series will work fine. If the cells are weak or the connections loose, perhaps 3 cells in series will be needed to make the bulb shine brightly.

2. Pulling on the steel wool in this manner will cause iron fibers to settle on the paper. Students should be careful not to get slivers in their fingers while they do this. Gloves or plastic bags, while not a requirement, do reduce the chances of getting slivers. A simple layer of masking tape on the pads of the fingers is also protective. As before, be sure your medical kit contains tweezers.

3. To observe the fireworks, students should place the pin and paper clip probes very close together, allowing the electricity to pass through short sections of the fine steel wool particles. They will watch in total fascination as the steel flashes and melts, sometimes burning tiny holes in the index card and causing puffs of smoke!

3c. Students have already learned that thin steel wool strands have higher electrical resistance than thicker ones. The important new concept here is that high resistance releases energy in the form of heat and light. To help students appreciate this in concrete terms, see the extension activity below.

4. The light bulb filament belongs at the top of the resistance list for two reasons: First, using direct observation, it appears to be very thin, even when compared to the fine steel wool fibers. Second, and more convincing, it releases a great deal of energy, mostly in the form of light, when electricity passes through it.

Extension

The heating element on a stove gets hot. A light bulb filament shines bright. Such increases in radiated energy associated with an increase in resistance can be illustrated by rubbing the palms of your hands together.

First rub them lightly; the resistance is low, so little heat is produced. Now rub them firmly together; the resistance is higher, so more heat is produced.

Answers

3A. The tiny fibers tend to flash and melt, without lighting the bulb.

3B. The steel wool fibers remain unchanged, and the bulb turns on.

3C. The thin strands have higher electrical resistance than the thick strands and clumps, so they heat up to a higher temperature, flash and melt. The thick strands and clumps, having lower resistance, allow the passage of enough electricity to light the bulb.

4.
   1: bulb filament
   2: thinner steel strand
   3: thicker steel strand
   4: foil ribbon

Materials

☐ A straight pin.

☐ Circuit components: dry cell holder, bulb holder, foil ribbon and paper clips. Student should share to make at least 2 cells available per activity group.

☐ An index card, preferably 3x5 inches.

☐ A ball of fine grade steel wool.

☐ Finger protection (optional: see note 2 above).

SURPRISE CIRCUITS

1

1 bulb
1 cell } *in*
1 switch } *parallel*

Diagram this circuit . . .

2 Build the circuit. When you push the switch on, what is the surprise?

3 Explain how your circuit works.

USE THE IDEA OF RESISTANCE IN YOUR ANSWER.

4 Build this circuit:

CALL THE LEFT SWITCH "L" AND THE RIGHT SWITCH "R"

L   R

5 Use diagrams and arrows to show how electrons move through the circuit when you turn on . . .

SWITCH L ONLY

SWITCH R ONLY

BOTH SWITCHES L and R

6 What is the surprise when both switches are turned on? Use the idea of resistance to explain what you see.

TOPS LEARNING SYSTEMS

Objective

To appreciate that the flow of electricity decreases with increased resistance.

Lesson Notes

2. Students should not leave the switch pushed down for extended periods of time: this creates an energy-draining short circuit.

3. It is technically incorrect to say that *all* the electrons follow the path of least resistance through the closed switch. Ohm's law predicts that electricity will flow through *both* the bulb *and* the switch, inversely proportional to their respective resistances. Since the resistance of the switch relative to the bulb is very low, most (but not all) electrons travel the path of least resistance across the switch.

To illustrate this idea, try building this circuit.

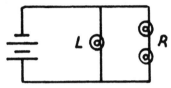

Most of the electrons flow through **L** because this is the path of least resistance. But the bulbs through **R** still shine feebly. Because **R** has twice the resistance, Ohm's law predicts that it will still receive half as much electricity as **L**. So electricity does "follow the path of least resistance", but it's *not* an "all or nothing" principle.

6. It is instructive to redraw the circuit like this.

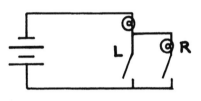

The bulb and switch through **R** have much higher resistance than the switch through **L**. Ohm's law predicts that most (not all) of the electricity will flow through **L**, the path of least resistance.

Extension

Divide your class into groups with perhaps 4 students each. Have each group *design* (but not build) a supercircuit. They can use any number of bulbs, switches and cells, up to the total number possessed by the group (4 each).

Each group must make certain that there are no short circuits when the switches are open, and that no bulb is exposed to more than 4.5 volts (3 cells in a series). If the teacher or any group member discovers a short, or the possibility of burning out a bulb, that circuit must be redesigned.

Each group then *builds* the circuit they have designed, and writes a report describing its properties. You select the most interesting circuits as a basis for further class demonstrations and discussions.

Shown here is one example of a supercircuit built with

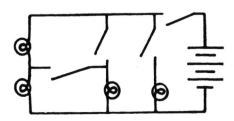

equipment pooled by four students. As you can see, the circuits may be very complicated. Be prepared for lots of questions,

some that you and your students may not be able to answer. Knowing it all, of course, is not the point. The point is to allow your students to experience science as a process of inquiry: to question, argue, hypothesize, experiment and observe; to experience the satisfaction of being right; to wonder why they are wrong.

Answers

1.

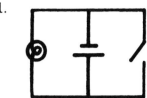

2. The switch is backward! When I push the switch on, the light goes off. When I release the switch to turn it off, the light goes on.

3. When I turn the switch on, an alternate low resistance path is provided. Most of the electrons follow this path of least resistance, by-passing the bulb.

5.

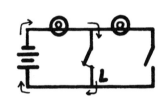

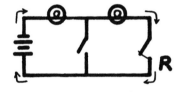

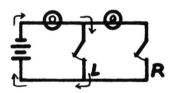

6. When both switches are turned on, only one bulb lights. Most of the electrons follow path L throught just one bulb, because it is the path of least resistance.

Materials

☐ Circuit components: dry cell holder, bulb holder, switch, foil ribbon and paper clips. Students should share equipment to make at least 2 cells, 2 bulbs and 2 switches available per activity group.

BUILD A FUSE

1 Cut out 1/4 of an index card.

2 Pull off a single strand of steel wool as thick as a hair, and tape each end to the piece of card.

3 Clamp 2 pennies on the steel wool with clothespins. Keep the space between them very small.

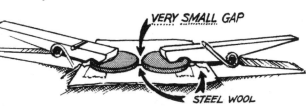

VERY SMALL GAP

STEEL WOOL

4 Clamp a foil ribbon over each penny. Keep the pennies so close that they almost touch.

NOW IT'S A FUSE!

5 Put this fuse in series with 2 dry cells and a bulb: The bulb should shine brightly.

ALL the electrons flow through the steel wool.

If the fuse MELTS, make one with a WIDER gap!

6 While the bulb is shining brightly, by-pass it with another ribbon. Keep your eye on the fuse!

BY-PASS

If the fuse WON'T melt, make one with a NARROWER gap!

7 Electric wires in a house can get too **HOT** if . . .

the appliances are by-passed (SHORT CIRCUIT) . . .

too many appliances draw electricity off the same line (OVERLOAD).

How does a fuse protect a house from fire?

A fuse should be neither too weak nor too strong. Explain.

TOPS LEARNING SYSTEMS

Objective

To understand how fuses work to protect circuits from shorts and overloads.

Lesson Notes

1. Only ¼ of the index card is needed per student. The other 3 parts can be used by others.

2. To make a fuse, the finer steel wool fibers work best. Fibers that look thicker than hair will probably not melt.

Be sure your students tape the steel wool only at the *ends* of the card. The middle must remain free of tape.

3-4. Be sure *both* pennies press against the steel wool. For the electricity to pass through the fuse, it must flow from one penny to the other via the strand.

It is crucial to keep this gap narrow, about the thickness of a dime. If it is any wider, the fuse will not melt when the circuit shorts out in step 6.

5-6. The bulb must shine brightly in step 5, so that when it's by-passed in step 6, there will be a sufficient surge of electricity to melt the fuse.

Your students may have to try several times before making a fuse that works. The trick is to make the fuse strong enough to sustain the current through the light bulb, but weak enough to melt when the light shorts out. They can control the strength of the fuse by adjusting the width of the gap across the two pennies. (The thickness of the steel wool fiber should remain constant: always choose ones that look about as thick as a hair.)

Gap is too wide. Bulb will not shine brightly, nor will the fuse melt.

Gap is still too wide. Bulb shines brightly, but the fuse won't melt when it shorts out.

THIN AS A DIME

Gap is **JUST RIGHT**. Bulb shines brightly and the fuse melts when it shorts out.

Gap is too narrow. The fuse melts immediately, before the bulb shorts out.

Extension

Inspect a fuse box in your school building. Does it contain fuses or circuit breakers? Find out how circuit breakers are different from fuses.

Look under the hood of your car (or perhaps under the dashboard near the steering column). Locate the fuses. What do they look like?

Answers

7. A fuse protects a house from fire because it is the weakest link in a circuit. If too much electricity flows through the line because of a short circuit or overload, the fuse is the first to heat up and melt, before the other wires can get hot enough to catch the house on fire.

If the fuse is too weak, it will melt whenever electricity flows through the circuit, whether or not there is an overload or short circuit. If the fuse is too strong, it won't melt in time to protect other wires from overheating.

Materials

☐ An index card, preferably 3x5 inches.

☐ Pair of scissors.

☐ Steel wool.

☐ Tape.

☐ Pennies.

☐ Clothespins.

☐ Circuit components: dry cell holder, bulb holder, foil ribbons and paper clips. Students should share to make at least 2 cells available per activity group.

BIG BANG!

1 Tape a strand of steel wool to a balloon.

2 Lay 2 foil ribbons, end to end, along the strand so the ends almost touch. Tape across the gap.

THE GAP IS AS THIN AS A DIME!

3 Connect the ribbons to an open switch and 2 dry cells.

Now close the switch . . .

4 Number these boxes in the proper order to explain why the balloon pops.

☐ *The hot wire melts the skin of the balloon.*

☐ *Electricity passes through the very thin steel strand.*

☐ *Electricity flows from the dry cell.*

☐ *The balloon pops!*

☐ *The thin wire has high resistance, so it gets hot.*

5 Use your own words to explain why the balloon pops. Write a paragraph.

TOPS LEARNING SYSTEMS

Objective

To explode a balloon with electricity. To understand the role of electrical resistance in this process.

Lesson Notes

1. Students should select one of the finer strands from the steel wool ball, about the thickness of a human hair.

2. The balloon is more likely to pop if the gap between the ribbons is small. It is crucial, however, to keep the ribbons from touching. If they do, electricity will not pass through the steel wool strand at all.

3. When the switch is pressed on, there are three possible events:

a. THE BALLOON POPS. Your students will almost certainly want to try the experiment several times over.

b. THE BALLOON DEFLATES RAPIDLY. Sometimes a small hole burns into the balloon's skin without breaking it. This is unexpected and quite humorous. When they try it again, your students should inflate the balloon with more air.

c. NOTHING HAPPENS. Check to see that the foil ribbons are taped securely over the steel strand, which must be touching the balloon. The ribbons should almost, but not quite, touch. Be certain that all connections are secure. Verify that the cells are placed in series, not opposition. If the balloon still won't pop, try adding a third or fourth cell in series. This should do the job. Normally a single cell is sufficient to break the balloon skin.

4-5. The paragraph in step 5 is easy to compose using the logic chain in step 4 as a model. Accept nothing less than a coherent, well-written paragraph in the student's own words.

Extension

Thread the Needle

a. Make a "needle" and "thread" by bending two paper clips.

b. Attach the needle and thread to opposite ends of 2 dry cells and a bulb holder, connected in series as shown.

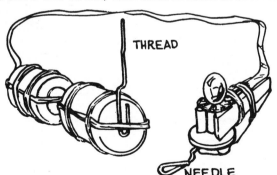

c. The challenge is to maneuver the needle all the way down the thread, and back up again, without lighting the bulb. The task can be made more (or less) difficult by making the "eye" of the needle smaller (or larger).

Answers

4. **4** The hot wire melts the skin of the balloon.

2 Electricity passes through the very thin steel strand.

1 Electricity flows from the dry cell.

5 The balloon pops!

3 The thin wire has high resistance, so it gets hot.

5. Close the switch, and electricity flows through the circuit. The steel wool, being so thin, has enough resistance to heat up and melt the skin of the balloon that presses against it. This pops the balloon.

Materials

☐ Balloons.

☐ Steel wool.

☐ Tape.

☐ Circuit components: dry cell holder, switch, foil ribbon and paper clips. Students should share to make at least 2 cells available per activity group.

2-WAY SWITCHES

1 Cut out 4 pieces of foil the size of index cards.

Divide each foil the long way to make 8 foil strips.

2 Wrap and tape the ends of 2 clothespins in the foil strips. Most of the foil should stick out past the ends.

TAPE HERE

3 Wrap all 4 ends of each clothespin. Make sure no foil touches the spring in the middle.

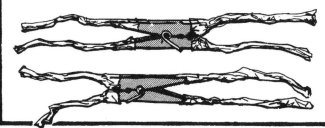

4 *Gently* twist the foil strips together like this. Be sure the switches point opposite ways.

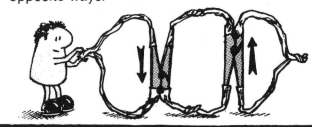

5 Connect your 2-way switches in series with a bulb and 2 dry cells.

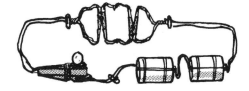

Why would 2-way switches like these be good to use at the top and bottom of a stairway?

6 Draw your circuit.

USE THIS SYMBOL FOR A 2-WAY SWITCH

7 Draw arrows to show 2 "on" positions and 2 "off" positions.

ON		OFF	
ON		OFF	

8 ***Have a contest.***
Ask a friend at one switch to keep the light turned off, while you try to keep it turned on at the other end.

TOPS LEARNING SYSTEMS

Objective

To build 2-way switches and intergrate them into a circuit. To understand how they work.

Lesson Notes

1. Index cards are mentioned only to give your students a rough idea as to how large they should cut the aluminum foil. Some may trace the exact outline of the card on the foil and then cut around it. This is OK, but unnecessary.

2-3. The foil must not touch the spring of either clothespin. If it does, a short circuit through the spring may develop in the 2-way switch.

4. The watchword here is *gently*. Your students are using foil that is not strengthened with a tape backing. It is easy to twist in two.

5. Two simple switches won't do the job. Connected in series, they can turn the lights on or off independently only if the other switch is left in an "on" position. Connected in parallel, both switches would have to be off to turn out the light.

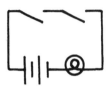 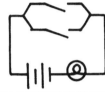

SIMPLE SWITCHES IN SERIES   **SIMPLE SWITCHES IN PARALLEL**

8. This little game of you-keep-it-on and I-keep-it-off is a favorite for kids to play in the home as long as no adults are yelling at them for wearing out the light switches. Here your students can play the game openly, wearing out their fingers before they ever break the clothespins.

But they must play fair. It is cheating to squeeze the clothespin to a midway position where the switch is neither on nor off. This breaks the circuit with no possibility of the other player reestablishing contact at the opposite end.

Answers

5. When 2-way switches are used at the top and bottom of stairs, either switch can be used to operate the same light. It doesn't matter whether the other switch is turned on or off.

6.

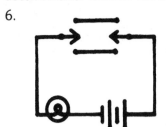

7.

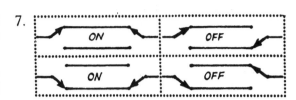

Materials

☐ Index card.

☐ Aluminum foil.

☐ Pair of scissors.

☐ Clothespins.

☐ Tape.

☐ Circuit components: dry cell holder, bulb holder, foil ribbon and paper clips. Students should share to make at least 2 cells available per activity group.

BULBS AND A PENNY

1 Connect 2 dry cells in series. Attach a foil ribbon and bulb to each end.

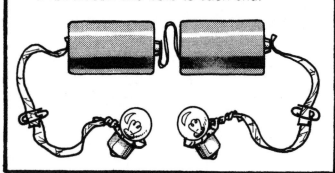

2 This is how the *filament* is connected to the *end* of the bulb and to the *collar*.

FILAMENT

3 DRAW LINES to show how electrons flow through each filament below. Tell if it shines *bright, dim, or not at all.*

A

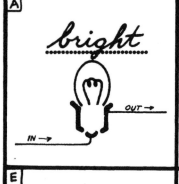

bright

IN →   OUT →

B

IN →   OUT →

C

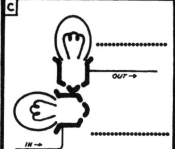

OUT →

IN →

D

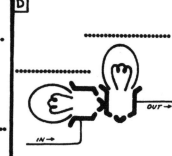

IN →   OUT →

E

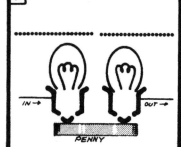

IN →   OUT →

PENNY

F

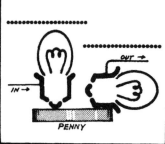

IN →   OUT →

PENNY

G

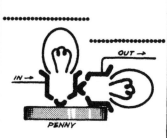

IN →   OUT →

PENNY

H

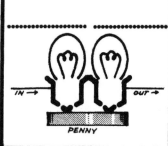

IN →   OUT →

PENNY

4 Why does the bulb in box C shine brighter than the bulbs in box B?

5 Compare boxes E and G. Which is in series and which is in parallel? Explain.

TOPS LEARNING SYSTEMS

Objective

To trace the flow of electricity through pairs of light bulbs. To recognize unusual series and parallel connections.

Lesson Notes

1. Bulbs *without* holders are required. If you use this activity in sequence as the last scheduled lesson in the module, the bulb holders can be disassembled, the bulbs used separately, and then all the various parts returned to storage until you teach the module again.

It is important to use bulbs manufactured by the *same* company, with the **same** resistance rating. Otherwise the bulbs may not shine with equal brightness when they should.

2-3. Understanding how a bulb is wired in step 2 will enable your students to complete the diagrams in step 3. Notice that two kinds of responses are required for each box: observations on bulb brightness, **and** lines to show the path that electrons follow through the bulbs.

The first observation on brightness is given to set a standard of comparison for the rest of the boxes.

5. The brightness of the bulbs in **G** also suggests a parallel connection, because the bulbs in **E** shine relatively dimly. Your students can see this difference most readily if they use one cell in the circuit instead of two. With a single cell, the parallel bulbs shine normally, but the series bulbs barely glow at all.

3E. *dim* *dim*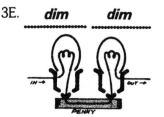

3F. *bright* *not at all*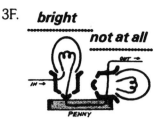

3G. *bright* *bright*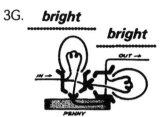

3H. *not at all* *not at all*

4. In box **B**, the bulbs share the electricity in series, so it passes through each bulb in turn. In **C**, the electricity passes only through the collar of one bulb to light the other. Being the only lighted bulb in the circuit, it shines brighter.

5. The bulbs in **G** are connected in parallel: each has an independent path to the cells, so that one can be disconnected without turning the other off. The bulbs in **E** are connected in series: they share the same path to the cells, so that if one is disconnected, they both turn off.

(The brightness of the bulbs in **G** also suggests a parallel connection, because the bulbs in **E** shine relatively dimly. Your students can see this difference most readily if they use *one* cell in the circuit instead of two. With a single cell, the parallel bulbs shine normally, but the series bulbs barely glow at all.)

Answers

3A. *bright*

3B. *dim* *dim*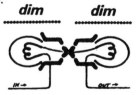

3C. *bright* *not at all*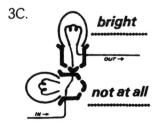

3D. *not at all* *bright*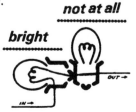

Materials

☐ Bulbs without holders. See note 1 above.

☐ A penny.

☐ Circuit components: dry cell holder, foil ribbons, and paper clips. Students should share to make a least 2 cells available per activity group.

SUPPLEMENTARY
PAGES

ELECTRO-SQUARES

supplement to Electro-Squares activity

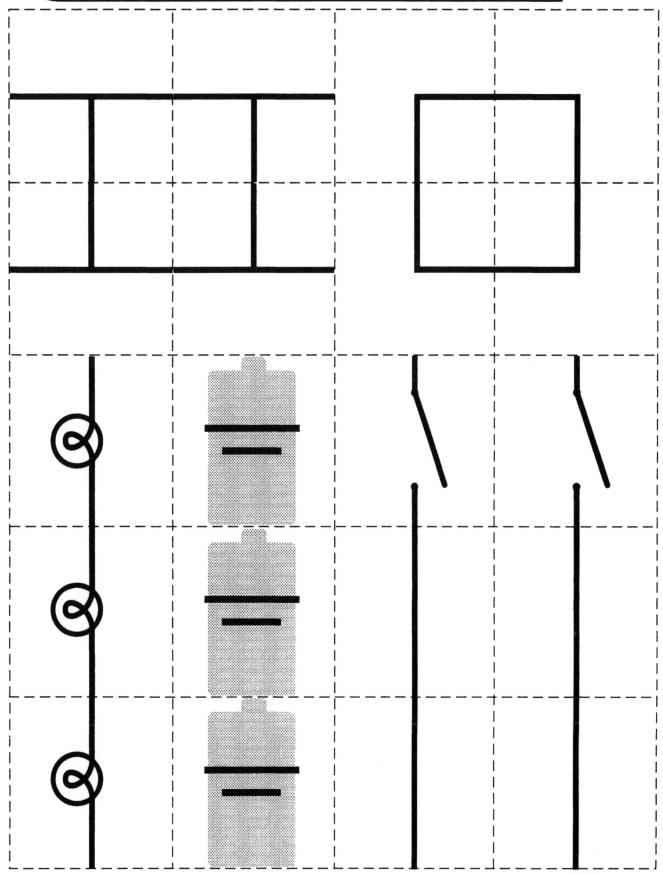

So, did you like ELECTRICITY?

YEAH! It was TOPS!

Let's do MAGNETISM next, or BALANCING, or RADISHES! Can we?

I LOVE science I can do myself!

YOU'LL LOVE THESE TOPS TITLES, TOO!

TASK CARD books: *open-ended activityes encourage scientific inquiry.*

01 **PENDULUMS** for grades 8-12
02 **MEASURING LENGTH** for grades 6-10
03 **GRAPHING** for grades 6-10
04 **BALANCING** for grades 6-11
05 **WEIGHING** for grades 5-10
06 **METRIC MEASURE** for grades 8-12
07 **MATH LAB** for grades 7-12
08 **PROBABILITY** for grades 6-10
09 **FLOATING & SINKING** for grades 7-12
10 **ANALYSIS** for grades 5-10
11 **OXIDATION** for grades 6-10
12 **SOLUTIONS** for grades 6-10
13 **COHESION/ADHESION** for grades 6-10
14 **KINETIC MODEL** for grades 7-12
15 **HEAT** for grades 8-12
16 **PRESSURE** for grades 7-12
17 **LIGHT** for grades 6-11
18 **SOUND** for grades 7-12
19 **ELECTRICITY** for grades 8-12
20 **MAGNETISM** for grades 8-12

21 **MOTION** for grades 7-12
22 **MACHINES** for grades 7-12
23 **ROCKS & MINERALS** for grades 6-12

ACTIVITY SHEET books: *a bit more step-by-step guidance for younger folks..*

31 **BALANCING** for grades 2-7
32 **ELECTRICITY** for grades 3-8
33 **MAGNETISM** for grades 3-8
34 **PENDULUMS** for grades 4-9
35 **METRIC MEASURING** for grades 5-9
36 **MORE METRICS** for grades 6-10
37 **ANIMAL SURVIVAL** for grades 3-8
38 **RADISHES** for grades 3-8
39 **CORN & BEANS** for grades 4-12
40 **EARTH, MOON & SUN** for grades 7-12
41 **PLANETS & STARS** for grades 7-12
42 **FOCUS POCUS** for grades 5-10

MICROSCOPES: *cheap, portable, fun!*

TOPScopes! grades 3-12. Build your own, or buy pre-asembled 25X or 36X microscopes you can carry in a pocket! (Other magnifying lenses also available.)

JOB CARD books: *open-ended exploration with lots of math and science for lower grades. Kits are available!*

71 **Pri. LENTIL SCIENCE** for grades K-3
72 **Inter. LENTIL SCIENCE** for grades 3-6
73 **GET A GRIP Workstation** for grades K-6

COMPILATIONS

91 **GLOBAL TOPS Teacher Manual** 100 activity sheets for grades 3-10. Student reference books and cutout booklets also available.

95 **TRY THIS ON FOR SCIENCE** A TOPS sampler! One lesson adapted from each of our books (excluding Lentil Science). Grade range 3-12.

MASTER TEACHER books: *great resources to make your classroom teaching easier and more rewarding.*

61 **A SUMMER START** grades 1-8
62 **Intermediate ALPHABET SOUP** grades 4-8
63 **PEACEFUL PROCEDURES** grades 1-8
64 **Primary ALPHABET SOUP** grades 1-3

Visit **www.topscience.org** to download free sample labs for all our fantastic books!

Feedback

If you enjoyed teaching TOPS please tell us so. Your praise motivates us to work hard. If you found an error or can suggest ways to improve this module, we need to hear about that too. Your criticism will help us improve our next new edition. Would you like information about our other publications? Ask us to send you our latest catalog free of charge.

For whatever reason, we'd love to hear from you. We include this self-mailer for your convenience.

Sincerely,

Ron and Peg Marson
author and illustrator

Your Message Here:

Module Title _____ Date _____

Name _____ School _____

Address _____

City _____ State _____ Zip _____